Computational Statistics - Predicting the Future from Sample Data

Edited by Christos Volos

Published in London, United Kingdom

Computational Statistics - Predicting the Future from Sample Data
http://dx.doi.org/10.5772/intechopen.1006214
Edited by Christos Volos

Contributors
Christos Volos, Edward Lakatos, Igor Lugo, Martha G. Alatriste-Contreras, Muhammad Aftab, Sthitadhi Das, Tanvir Ahmad

First published in London, United Kingdom, 2025 by IntechOpen
IntechOpen is the global imprint of INTECHOPEN LIMITED, registered in England and Wales, registration number: 11086078, 167-169 Great Portland Street, London, W1W 5PF, United Kingdom

For EU product safety concerns: IN TECH d.o.o., Prolaz Marije Krucifikse Kozulić 3, 51000 Rijeka, Croatia, info@intechopen.com or visit our website at intechopen.com.

British Library Cataloguing-in-Publication Data
A catalogue record for this book is available from the British Library

Computational Statistics - Predicting the Future from Sample Data
Edited by Christos Volos
p. cm.
Print ISBN 978-1-83634-617-3
Online ISBN 978-1-83634-616-6
eBook (PDF) ISBN 978-1-83634-618-0

If disposing of this product, please recycle the paper responsibly.

Meet the editor

Dr. Christos Volos is a Full Professor in the Laboratory of Nonlinear Systems, Circuits & Complexity (LaNSCom) of the Physics Department of the Aristotle University of Thessaloniki, Greece. His research interests include chaos, nonlinear systems, mem-elements, analog and mixed signal electronic circuits, neural networks, chaotic electronics and their applications (secure communications, cryptography, robotics), as well as chaotic synchronization and control. Dr. Volos has published more than 400 papers and chapters in international journals, scientific books, and international conferences, in addition to editing 10 books and 2 books as an author on the topic of nonlinear circuits and systems. Furthermore, he is an Editorial Board Member of 11 international journals and has been a Guest Editor of more than 30 Special Issues in international journals.

Contents

Preface

Predicting the future state of a system from sample data is a central concern of scientific research today. While deterministic mathematical models, such as those used in classical physics, have been known to provide accurate predictions under ideal conditions for centuries, the increasingly complex, noisy, and data-rich phenomena of the real world often defy such precise formulations. This is where computational statistics comes in, as a transformative tool that bridges the gap between abstract theory and empirical reality, allowing us to glean insights into the patterns underlying many systems and arrive at reliable predictions of their behavior.

This volume, entitled "*Computational Statistics – Predicting the Future from Sample Data*", has been edited with the aim of being a scientific response to the evolving role of computational statistics in our time. It includes five carefully curated chapters that demonstrate the breadth and depth of modern computational statistical methods and their interdisciplinary applications. From fundamental theory to specialized areas, this book aims to highlight how statistical science, combined with increasing computing power, can reveal the hidden structure in large volumes of data and enable their effective exploitation.

The first chapter, entitled "Introductory Chapter: Computational Statistics", offers a conceptual and historical overview of the specific field. It sets the stage by identifying how computational methods have reshaped traditional statistical tools and enabled the exploration of high-dimensional data, complex systems, and dynamic processes. Therefore, it sets the tone for the rest of the book, outlining the applications of computational statistics in various fields, as well as its implications for predicting behaviors and phenomena in today's uncertain world.

Chapter 2, "Kernel Smoothing in Semiparametric Regression", provides estimation techniques for the nonparametric regression function, including kernel smoothing, spline smoothing and local linear, as well as polynomial smoothing. Also, the estimation of the parametric explanatory part of the response is presented using a well-known technique introduced by Robinson in 1988. The asymptotic properties of the estimators of the parametric and nonparametric regression functions are also discussed to furnish a consistent prediction of the response.

Chapter 3, "Statistical Regularities in the Musical Work of Marin Marais, Pièces de Viole Des Cinq Livres", presents the analysis of the spectrum of audio signals related to the work of "Pièces de viole des Cinq Livres" based on the music of Marin Marais, performed by Jordi Savall. In particular, authors have tried to identify the statistical regularity underlying this musical work. Based on the complex systems approach, they compute the spectrum of audio signals and analyze and identify their best-fit statistical distributions.

Chapter 4, "Projecting Event Accrual as a Survival Trial Progresses", delves into a crucial aspect of finding methods for projecting the accrual of events for survival trials. The projection methods are based on the Markov model for survival trials, which was initially developed for calculating sample size and power for survival trials, adjusting for complexities routinely encountered in these settings. The factors that typically complicate these calculations involve time-dependent rates of all parameters for failure, loss-to-follow-up, loss to competing risks, non-compliance, non-constant treatment effects, and staggered entry. Therefore, it was explained why entry staggered differs from other complexities, and a detailed account of how it is modeled is presented.

The final chapter, "Future Prediction through Planned Experiments", presents a comprehensive examination of how experimental data can be used to make future predictions. Through a combination of theoretical concepts and practical examples, readers will gain a sound understanding of the predictive process for reliable decision-making and policy-making in real-world scenarios.

Together, these chapters form a coherent and highly engaging exploration of computational statistics and its applications in practice. While each chapter presents a distinct topic, they are all united by a common goal: that of using data through statistical methods and computational tools to better understand and predict the world around us. Therefore, whether the application is scientific, artistic, clinical, or experimental, the tools of computational statistics remain essential for transforming past observations and data into future knowledge.

We hope that this book can serve as an updated and handy reference for a broad audience, including statisticians, data scientists, university professors, graduate students, and researchers interested in deepening their understanding of how computational tools enhance statistical methods.

Christos Volos
Laboratory of Nonlinear Systems, Circuits & Complexity (LanSCom),
Physics Department,
Aristotle University of Thessaloniki,
Thessaloniki, Greece

Chapter 1

Introductory Chapter: Computational Statistics

Christos Volos

1. Introduction

Throughout the history of science, the combination of mathematics and empirical research has produced some of humanity's most remarkable achievements. From the sky observation and the prediction of celestial mechanics to the use of control systems in engineering, the predictive power of mathematics has shaped modern way of life and therefore our civilization. However, as we left behind the realm of physical laws and we move forward to that of complex systems, such as biological systems, financial markets, and social networks, we encounter a profound shift in our view. The determinism of Newtonian mechanics gives his place to noise, variability, and uncertainty. This is where statistics, and more recently computational statistics, emerges as an indispensable lens through which we understand the world.

Therefore, this book aims to explore this new research framework that is taking shape, according to which data is now closely linked to algorithms and prediction is no longer based on known physical laws, but on the non-obvious structures, hidden within the data. This is also indicated by the subtitle of the book, as it is not a simple search in science today, but a basic research methodology, followed in all natural, economic, and social sciences, as well as in commerce and industry.

2. From mathematical modeling to statistical prediction

The distinction between these two concepts, the mathematical modeling and statistical prediction, is not simply technical. It defines a contemporary trend and a deeper epistemological shift. Mathematical modeling, as used for example in classical mechanics, arises on the basis of fundamental principles of physics, such as symmetries and laws of conservation of energy or momentum, from which accurate predictions result. However, in 1960 Eugene Wigner mentioned that the success of this approach is both astonishing and limited [1]. He mentioned this, because it works very well in areas governed by physical laws. However, its scope of application is dramatically limited when applied to complex, multidimensional, or stochastic systems.

On the other hand, statistics does not require prior knowledge of the system under study and the rules and laws that govern its operation. Instead, it starts with data, often large, messy, and incomplete, and seeks to extract patterns that can form reasonable predictions. In this way, statistics tries to approach the variability of many systems and offer probabilistic rather than deterministic predictions [2]. Moreover, this approach has gained ground in recent decades due to the rapid increase

IntechOpen

in computing power, the design of new algorithms adapted to each problem, and the development of machine learning [3]. For the aforementioned reasons, computational statistics is today a constantly evolving field of research and practical application in data science.

3. The rise of computational statistics

However, what is the reason for the rapid development of computational statistics? The answer to this question is obviously related to the digital revolution that societies have experienced in recent decades. It is now a fact that sensors, networks, and storage technologies have evolved to such an extent, thanks to the availability of big data and information. From genetic sequences in Medicine, to climate models in Meteorology, and from interactions in social media to financial transactions, the number and complexity of available data surpasses traditional data analysis procedures.

Therefore, computational statistics can respond to these challenges in various ways:

- *Algorithmic efficiency*: New, more sophisticated algorithms now allow us to analyze big data. This includes algorithms adapted from classical procedures, such as regression, as well as advanced methods, such as stochastic gradient, expectation-maximization, and Markov Chain Monte Carlo [4].

- *Simulation-based inference*: In problems where analytical solutions are intractable, simulation techniques, such as bootstrap methods [5], permutation tests, and Bayesian posterior sampling, are applied, which allow for reliable inference.

- *Model flexibility*: Nonparametric and semiparametric models, supported by computational methods, have been developed, allowing researchers to model complex phenomena without assuming overly simplistic functions [6].

- *Data-driven discovery*: Computational statistics can play an important role in areas where theoretical models are inadequate or incomplete, because it can reveal empirical rules that will serve as predictive tools. This is already happening in fields, such as personalized medicine [7], and in real-time anomaly detection systems.

- *Integration with machine learning*: In recent years, the boundaries between machine learning and computational statistics have become increasingly blurred. Many predictive tools that have come to the fore, such as random forests, support vector machines, and neural networks, are both machine learning algorithms and statistical models [8–10].

The aforementioned developments have dramatically changed the way we think about achieving the goal of prediction. That is, instead of seeking to extract predictions from known equations, we often try to extract predictions from patterns hidden in the data. This rapid change in the way we think and work is not only practical, but also highlights the role that information plays in science today.

4. Applications across disciplines

It is now a fact that computational statistics has an interdisciplinary scope and is involved in many areas of research. While we know that traditional statistical methods were often limited to specific areas, such as in agriculture or survival analysis in medicine, computational statistics has expanded its application to almost every scientific field. In more details:

- *In the life sciences*, it has achieved significant results through genomics, proteomics, and systems biology using high-dimensional inference and network analysis [11].

- *In economics and finance*, it has played an important role, especially in the last two decades, as it feeds algorithmic trading, risk modeling, and credit rating with real-time data [12].

- *In the social sciences*, it has made significant contributions to modeling human behavior, political opinions, and information dissemination in social networks [13].

- *In the physical sciences*, it has provided significant assistance in the analysis of complex simulations, in the calibration of models with observational data, and in the quantification of uncertainties in prediction problems [1].

In summary, we can say that what unites the aforementioned applications is their common basis, which is statistics. Through this, conclusions can be drawn, satisfactory prediction can be achieved, and uncertainty can be quantified. However, an important factor enhancing the effectiveness of statistics today is the large amount of available computing power and the improvement of algorithmic complexity.

5. Challenges and opportunities

However, computational statistics, in addition to the new horizons it opens up in research, is also accompanied by significant challenges. As we delegate more and more decision-making to data-based systems in critical areas of modern societies, from health to economics and justice, issues such as the interpretability of decisions and the ethical use of these systems become crucial [14]. It is also a fact that the opacity that characterizes the design of some deep neural networks, which are used in prediction problems, has raised discussions about the degree of trust we can have in them.

Furthermore, it is well known that the abundance of data that overwhelms us daily does not automatically yield the desired knowledge. Learning from data through statistical tools requires special attention to critical issues, such as sampling bias, overfitting, and spurious correlations [15]. Even the most sophisticated computational techniques, without a rigorous statistical basis, can still lead to erroneous conclusions.

Therefore, the future of computational statistics lies not only in designing better algorithms, but also in developing a rationale for better understanding the principles that govern statistics as a mechanism for collecting and processing data. This includes classic concepts such as estimation, hypothesis testing, and confidence intervals, as well as newer ideas such as normalization, cross-validation, and Bayesian decision theory.

6. Conclusion

Computational statistics represents a major advance in statistical analysis, providing an expanded and more flexible framework for modeling complex systems. In a world increasingly shaped by large amounts of data, the ability to predict a future state of a system from sample data is more than a technical skill. It is a scientific need that the research community must respond to. Therefore, computational statistics offers the tools to meet this challenge.

Author details

Christos Volos
Laboratory of Nonlinear Systems, Circuits and Complexity Department of Physics, Aristotle University of Thessaloniki, Greece

*Address all correspondence to: volos@physics.auth.gr

References

[1] Russ S. The unreasonable effectiveness of mathematics in the natural sciences. Interdisciplinary Science Reviews. 2011;**36**(3):209-213

[2] Casella G, Berger R. Statistical Inference. Florida, USA: CRC Press; 2024

[3] Hastie T, Tibshirani R, Friedman J. The Elements of Statistical Learning: Data Mining, Inference, and Prediction. New York, USA: Springer; 2009

[4] Robert CP, Casella G. Monte Carlo Statistical Methods. New York, USA: Springer; 1999

[5] Efron B, Tibshirani RJ. An Introduction to the Bootstrap. New York, USA: Chapman & Hall; 1993

[6] Wasserman L. All of Nonparametric Statistics. Berlin, Germany: Springer; 2006

[7] Collins FS, Varmus H. A new initiative on precision medicine. New England Journal of Medicine. 2015;**372**(9):793-795

[8] Murphy KP. Machine Learning: A Probabilistic Perspective. Cambridge, USA: MIT Press; 2012

[9] Breiman L. Random forests. Machine Learning. 2001;**45**(1):5-32

[10] Kennedy MC, O'Hagan A. Bayesian calibration of computer models. Journal of the Royal Statistical Society: Series B (Statistical Methodology). 2001;**63**(3):425-464

[11] Storey JD, Tibshirani R. Statistical significance for genomewide studies. PNAS. 2003;**100**(16):9440-9445

[12] Tsay RS. Analysis of Financial Time Series. New Jersey, USA: Wiley; 2005

[13] Kruschke JK. Bayesian data analysis. Wiley Interdisciplinary Reviews: Cognitive Science. 2010;**1**(5):658-676

[14] Pessach D, Shmueli E. A review on fairness in machine learning. ACM Computing Surveys. 2022;**55**(3):1-44

[15] Domingos PA. Few useful things to know about machine learning. Communications of the ACM. 2012;**55**(10):78-87

Chapter 2

Kernel Smoothing in Semiparametric Regression

Sthitadhi Das

Abstract

A semiparametric regression model consists of parametric explanatory part of the response as well as nonparametric regression function of one or more variable(s) interpreting the response. The basic semiparametric regression model involves a linear function of a single parametric covariate as well as an unknown but preferably nonlinear function of a single nonparametric covariate. The scope of this chapter is to provide estimation techniques for the nonparametric regression function, including kernel smoothing, spline smoothing and local linear as well as polynomial smoothing. Also, the estimation of the parametric explanatory part of the response can be done using the technique by *Robinson* (1988). The asymptotic properties of the estimators of the parametric and nonparametric regression functions need to be discussed to furnish a consistent prediction of the response.

Keywords: kernel density estimation, semiparametric regression, spline smoothing, local linear smoothing, local polynomial smoothing

1. Introduction

In the field of Statistics, the wide application of the semiparametric regression model is significantly noteworthy. In regression theory, the basic approach to studying a variable of interest is to consider a linear function of one independent variable, usually termed a covariate, or several independent (or explanatory) variables/covariates/concomitants. If we want to analyze the scores of 40 students in a class in a Biology examination on the basis of the concomitant "study hour", preferably a linear function of it can furnish the objective. Denoting "study hour" by X, a simple linear regression model can be formed as $Y = \alpha_0 + \alpha_1 X$ where Y indicates the scores of the students in Biology, which further provides estimates of β_0, β_1 obtained through traditional least squares estimation or any other convenient technique(s). A multiple linear regression model can be achieved when some more independent variables are included in the model to delineate the scores, for example, students' mental health status, their previous results, and involvement in extracurricular activities. In that case, a model with the mathematical representation $Y = \beta_0 + \beta_1 X_1 + \ldots + \beta_p X_p + \epsilon$ is a suitable one, with ϵ being the lack in depicting the study variable of interest. The quantities $\beta_0, \ldots, \beta_p$ are known as parameter(s) of the regression model; hence, the whole setup is often addressed as a parametric regression setup.

Other than any linear function of the explanatory variable(s), a nonlinear type regression function might be an appropriate choice in various cases. For example, the displacement of a car from its initial position to its final position with more or less uniform acceleration is a parabolic function of the time taken in the course, that is, $S \approx at + bt^2$ where a, b are the parameters. In fact, $Y = a\log(X_1) + be^{X_2} + \epsilon$ or a model with p degree polynomial predictor $Y = \beta_0 + \beta_1 X + \beta_2 X^2 + \dots + \beta_p X^p$ are example of nonlinear regression models. In these models, the parameters can be estimated through various efficient estimation procedures [1–12].

A more general thought on studying the response variable Y is to remove the assumption of linearity or non-linearity of the function of concomitants, that is, an unknown or unspecified function of the independent variables would lead us to predict on the response variable eventually. One can consider the model $Y = f(X) + \epsilon$ or $Y = f(X_1, \dots, X_l) + \epsilon$ based on l covariates $X_1, \dots, X_l$ where the mathematical structures of the regression functions $f(X)$ or $f(X_1, \dots, X_l)$ are not known apriori. Such a model is labeled as *nonparametric regression model*. Yet, it is possible to estimate $f(X)$ or $f(X_1, \dots, X_l)$ using traditional kernel density estimation methods suggested by Nadaraya-Watson known as *Nadaraya-Watson kernel density estimation method* or local linear/polynomial smoothing technique or spline smoothing method or k-nearest neighbor method etc. By adding some linear/nonlinear predictor to the nonparametric regression function in the concerning model under investigation, one can get the function of independent variables as a mixture of parametric and nonparametric explanatory components to predict on the study variable. This type of model is called *semiparametric regression model*, for exampe, $Y = \beta_0 + \beta_1 X + f(V)$ where $\beta_0 + \beta_1 X$ is the linear parametric function of the covariate X and $f(V)$ is the nonparametric function of another covariate V. The model is also known as *partially linear regression model*, a typical semiparametric regression model which is frequently used in various fields including finance, business, medical science, economics etc. Also, $Y = a\log(X_1) + be^{X_2} + f(W_1, W_2, W_3)$ is another type of semiparametric regression model with the nonlinear parametric predictor $a\log(X_1) + be^{X_2}$ of X_1, X_2 and $f(\cdot, \cdot, \cdot)$ is an untractable function of three covariates W_1, W_2, W_3.

2. Kernel density estimation in semiparametric regression

A usual semiparametric regression model is expressed as a combination of some deterministic function of parameters as well as an unknown function of several useful concomitants. A general representation of such a model is $Y = \psi(X_1, \dots, X_u; \beta_1, \dots, \beta_u) + h(Z_1, \dots, Z_v) + \epsilon$ where Y is the response studied by parametric concomitants $X_1, \dots, X_u$ and nonparametric covariates $Z_1, \dots, Z_v$ through proper functions. The approach of non-parametric kernel smoothing is quite desirable in such cases for predicting Y feasibly. When $\psi(X_1, \dots, X_u; \beta_1, \dots, \beta_u)$ is a linear function of parameters $\beta_1, \dots, \beta_u$, then the estimation of $\beta_1, \dots, \beta_u$ is done first by applying the method of *Robinson (1988)* [3]. For a nonlinear function $\psi(X_1, \dots, X_u; \beta_1, \dots, \beta_u)$ of the parameters, some intuitive approaches can be made further. Let us discuss the relevant methodologies in both the setups of partially linear and partially nonlinear models as follows.

2.1 Kernel density estimation in a partially linear model

The model $Y = \psi(X_1, \dots, X_u; \beta_1, \dots, \beta_u) + h(Z_1, \dots, Z_v) + \epsilon$ denotes a partially linear model when $\psi(X_1, \dots, X_u; \beta_1, \dots, \beta_u) = \beta_1 X_1 + \dots + \beta_u X_u$, a linear function of the

u parametric covariates $X_1, \dots, X_u$. Here, the intercept of the model is merged with random error ϵ. So, the procedure of estimation of $\beta_1, \dots, \beta_u$ is to be furnished initially, with the help of *Robinson* (1988) [3]'s approach. The stepping stone of this entire procedure is the assumption on ϵ, *viz.* $E(\epsilon|X_1, \dots, X_u, Z_1, \dots, Z_v) = 0$, for all values of the $(u + v)$ covariates under consideration, $E(\epsilon^2|X_1, \dots, X_u, Z_1, \dots, Z_v) = \theta(X_1, \dots, X_u, Z_1, \dots, Z_v)$ which is a finite positive quantity. The estimation technique is accomplished in the following manner:

$$E(Y|Z_1, \dots, Z_v) = \sum_{i=1}^{u} \beta_i E(X_i|Z_1, \dots, Z_v) + E(h(Z_1, \dots, Z_v)|Z_1, \dots, Z_v) + E(\epsilon|Z_1, \dots, Z_v) \quad (1)$$

where
$E(h(Z_1, \dots, Z_v)|Z_1, \dots, Z_v) = h(Z_1, \dots, Z_v)$ and
$E(\epsilon|Z_1, \dots, Z_v) = E_{X_1,\dots,X_u}(E(\epsilon|X_1, \dots, X_u, Z_1, \dots, Z_v)) = 0$. Let us denote $E(Y|Z_1 = z_1, \dots, Z_v = z_v) = h_Y\left(\underset{\sim}{z}\right)$ where $\underset{\sim}{z} = (z_1, \dots, z_v)'$ is a given value of $\underset{\sim}{Z} = (Z_1, \dots, Z_v)'$. Then we can further write

$$h_Y\left(\underset{\sim}{z}\right) = \sum_{i=1}^{u} \beta_i h_{X_i}\left(\underset{\sim}{z}\right) + h\left(\underset{\sim}{z}\right) \quad (2)$$

where $h_{X_i}\left(\underset{\sim}{z}\right) = E(X_i|Z_1 = z_1, \dots, Z_v = z_v), i = 1, \dots, u$. Then, we subtract this model from the original model $Y = \beta_1 X_1 + \dots + \beta_u X_u + h(Z_1, \dots, Z_v) + \epsilon$ and get a transformed regression model

$$Y - h_Y\left(\underset{\sim}{z}\right) = \sum_{i=1}^{u} \beta_i \left(X_i - h_{X_i}\left(\underset{\sim}{z}\right)\right) + \epsilon \quad (3)$$

that is, $\epsilon_{Y\underset{\sim}{Z}} = \underset{\sim}{\beta}^T \underset{\sim}{\epsilon}_{X\underset{\sim}{Z}} + \epsilon$ where $\underset{\sim}{\epsilon}_{X\underset{\sim}{Z}} = \left(X_1 - h_{X_1}\left(\underset{\sim}{Z}\right), \dots, X_u - h_{X_u}\left(\underset{\sim}{Z}\right)\right)'$. This model (3) is indeed a simple linear regression model, and the traditional least squares method of estimation of $\underset{\sim}{\beta}$ is done and its estimator is obtained as

$$\underset{\sim}{\hat{\beta}} = \left(\underset{\sim}{\epsilon}^T_{X\underset{\sim}{Z}} \underset{\sim}{\epsilon}_{X\underset{\sim}{Z}}\right)^{-} \left(\underset{\sim}{\epsilon}^T_{X\underset{\sim}{Z}} \underset{\sim}{\epsilon}_{X\underset{\sim}{Z}}\right) \quad (4)$$

where $\left(\underset{\sim}{\epsilon}^T_{X\underset{\sim}{Z}} \underset{\sim}{\epsilon}_{X\underset{\sim}{Z}}\right)^{-}$ denotes the generalized inverse (g-inverse) of $\left(\underset{\sim}{\epsilon}^T_{X\underset{\sim}{Z}} \underset{\sim}{\epsilon}_{X\underset{\sim}{Z}}\right)$, and if it is a full rank matrix of order u, then we apply the inverse of $\left(\underset{\sim}{\epsilon}^T_{X\underset{\sim}{Z}} \underset{\sim}{\epsilon}_{X\underset{\sim}{Z}}\right)$, that is, $\left(\underset{\sim}{\epsilon}^T_{X\underset{\sim}{Z}} \underset{\sim}{\epsilon}_{X\underset{\sim}{Z}}\right)^{-1}$ as a special case. Next, a transformation of Y is made as $Y' = Y - \underset{\sim}{\hat{\beta}}^T \underset{\sim}{X}$ and subsequently, the regression model under study is converted to

$$Y' = h(Z_1, \dots, Z_v) + \epsilon \quad (5)$$

which is a nonparametric regression model indeed. Estimation of $h(Z_1, \ldots, Z_v)$ can be furnished in several ways, among which a preferably convenient technique is the *kernel density estimation due to Nadaraya-Watson* in multivariate setup. At a given value of $\underset{\sim}{Z} = (Z_1, \ldots, Z_v)'$, the Nadaraya-Watson kernel density estimator is derived stepwise in the following manner [1, 13, 14]:

$$h\left(\underset{\sim}{Z}\right) = E\left(Y'|\underset{\sim}{Z}\right) = \int_{-\infty}^{\infty} y' \cdot \frac{f_{Y',\underset{\sim}{Z}}\left(y', \underset{\sim}{z}\right)}{g_{\underset{\sim}{Z}}\left(\underset{\sim}{z}\right)} dy' \tag{6}$$

where $f_{Y',\underset{\sim}{Z}}(\cdot, \cdot)$ is the joint probability density function of $\left(Y', \underset{\sim}{Z}\right)$ and $g_{\underset{\sim}{Z}}(\cdot)$ is the marginal probability density function of $\underset{\sim}{Z}$. Furthermore, the density functions are calculated as

$$f_{Y',\underset{\sim}{Z}}\left(y', \underset{\sim}{z}\right) \hat{=} \frac{1}{nd_{y'}\prod_{t=1}^{v} d_t} \sum_{i=1}^{n} k\left(\frac{y' - Y_i'}{d_{y'}}, \frac{z_1 - Z_{1i}}{d_1}, \ldots, \frac{z_v - Z_{vi}}{d_v}\right) \tag{7}$$

where $k(\cdot, \ldots, \cdot)$ denotes a $(v + 1)$ -dimensional kernel density function of $\left(Y', \underset{\sim}{Z}\right)$. It is further expressed as a product kernel as

$$k\left(\frac{y' - Y_i'}{d_{y'}}, \frac{z_1 - Z_{1i}}{d_1}, \ldots, \frac{z_v - Z_{vi}}{d_v}\right) = k_{Y'}\left(\frac{y' - Y_i'}{d_{y'}}\right) \times \prod_{j=1}^{v} k_{Z_j}\left(\frac{z_j - Z_{ji}}{d_j}\right), \tag{8}$$

$i = 1, \ldots, n$, with $k_{Y'(\cdot)}$ and $k_{Z_j}(\cdot)$ being the marginal kernel density functions of Y' and $Z_j, j = 1, \ldots, v$, respectively. Similarly, $g_{\underset{\sim}{Z}}\left(\underset{\sim}{z}\right)$ is estimated as

$$g_{\underset{\sim}{Z}}\left(\underset{\sim}{z}\right) \hat{=} \frac{1}{n\prod_{t=1}^{v} d_t} \sum_{i=1}^{n} k\left(\frac{z_1 - Z_{1i}}{d_1}, \ldots, \frac{z_v - Z_{vi}}{d_v}\right) = \frac{1}{n\prod_{t=1}^{v} d_t} \sum_{i=1}^{n} \left\{\prod_{j=1}^{v} k_{Z_j}\left(\frac{z_j - Z_{ji}}{d_j}\right)\right\}. \tag{9}$$

Here, the quantities $d_{y'}, d_1, \ldots, d_v(> 0)$ are the bandwidths of estimation of kernel density functions of $Y', Z_1, \ldots, Z_v$, respectively.

Finally, an estimator of $h\left(\underset{\sim}{z}\right)$ is obtained by formulating (6) as

$$\hat{h}\left(\underset{\sim}{z}\right) = \int_{-\infty}^{\infty} y' \cdot \frac{\frac{1}{nd_{y'}\prod_{t=1}^{v} d_t} \sum_{i=1}^{n} k\left(\frac{y' - Y_i'}{d_{y'}}, \frac{z_1 - Z_{1i}}{d_1}, \ldots, \frac{z_v - Z_{vi}}{d_v}\right)}{\frac{1}{n\prod_{t=1}^{v} d_t} \sum_{i=1}^{n} \left\{\prod_{j=1}^{v} k_{Z_j}\left(\frac{z_j - Z_{ji}}{d_j}\right)\right\}} dy' \tag{10}$$

$$= \int_{-\infty}^{\infty} \frac{y'}{d_{y'}} \cdot \frac{\sum_{i=1}^{n} k_{Y'}\left(\frac{y' - Y_i'}{d_{y'}}\right) \times \prod_{j=1}^{v} k_{Z_j}\left(\frac{z_j - Z_{ji}}{d_j}\right)}{\sum_{i=1}^{n} \left\{\prod_{j=1}^{v} k_{Z_j}\left(\frac{z_j - Z_{ji}}{d_j}\right)\right\}} dy' \tag{11}$$

$$= \int_{-\infty}^{\infty} \frac{y'}{d_{y'}} \cdot \sum_{i=1}^{n} \omega_i \cdot k_{Y'}\left(\frac{y' - Y'_i}{d_{y'}}\right) dy' \text{ where } \omega_i = \frac{\left\{\prod_{j=1}^{v} k_{Z_j}\left(\frac{z_j - Z_{ji}}{d_j}\right)\right\}}{\sum_{i=1}^{n}\left\{\prod_{j=1}^{v} k_{Z_j}\left(\frac{z_j - Z_{ji}}{d_j}\right)\right\}}, i = 1, \dots, n \tag{12}$$

$$= \sum_{i=1}^{n} \int_{-\infty}^{\infty} \omega_i \cdot \frac{y'}{d_{y'}} k_{Y'}\left(\frac{y' - Y'_i}{d_{y'}}\right) dy' \tag{13}$$

$$= \sum_{i=1}^{n} \int_{-\infty}^{\infty} \omega_i \cdot y' k_{Y'}\left(\frac{y' - Y'_i}{d_{y'}}\right) d\left(\frac{y' - Y'_i}{d_{y'}}\right) \tag{14}$$

$$= \sum_{i=1}^{n} \int_{-\infty}^{\infty} \omega_i \cdot \left(Y'_i + d_{y'} \cdot s_i\right) ds_i \text{ with } s_i = \frac{y' - Y'_i}{d_{y'}}, i = 1, \dots, n \tag{15}$$

$$= \sum_{i=1}^{n} \omega_i Y'_i \int_{-\infty}^{\infty} k_{Y'}(s_i) ds_i + d_{y'} \sum_{i=1}^{n} \omega_i \int_{-\infty}^{\infty} s_i k_{Y'}(s_i) ds_i \tag{16}$$

$$= \sum_{i=1}^{n} \omega_i Y'_i, \tag{17}$$

since by the definition of a kernel density function $\phi(\cdot)$ defined on $\mathbb{R}$, $\int_{-\infty}^{\infty} \phi(a) da = 1$ and $\int_{-\infty}^{\infty} a\phi(a) da = 0$.

In the above kernel density estimation scheme, the bandwidth matrix considered is $Diag\left(d_1^2, \dots, d_v^2\right) = \mathbf{M}, say, d_1, \dots, d_v > 0$. However, we are generally interested about optimal bandwidth matrix and the criterion for such consideration can be fulfilled by calculating AMISE (asymptotic mean integrated squared error) of **M** which is finally obtained as

$$AMISE(\mathbf{M}) = n^{-1}\left(\prod_{c=1}^{v} d_c\right)^{-1/2} \mathscr{R}(K) + \frac{1}{4}[m_2(K)]^2 [vec.\mathbf{M}]^T \Psi(vec.\mathbf{M}) \tag{18}$$

where $\mathscr{R}(K) = \int_{\mathbb{R}^v} \left[K\left(\underset{\sim}{z}\right)\right]^2 d\underset{\sim}{z}$, $\int_{\mathbb{R}^v} \underset{\sim}{z}\underset{\sim}{z}^T k\left(\underset{\sim}{z}\right) d\underset{\sim}{z} = m_2(K) I_v$ where I_v is $v \times v$ identity matrix and $\Psi = \int_{\mathbb{R}^v} \left(vec.D^2 h\left(\underset{\sim}{z}\right)\right)\left(vec.^T D^2 h\left(\underset{\sim}{z}\right)\right) d\underset{\sim}{z}$ with $vec.$ Being the traditional vector operator on a matrix that stacks all the columns of the matrix into a single vector form, and

$$D^2 h\left(\underset{\sim}{z}\right) = \left(\left(\frac{\delta^2 h\left(\underset{\sim}{z}\right)}{\delta z_i \delta z_j}\right)\right)_{i,j=1,\dots,v} \tag{19}$$

is the Hessian matrix of order v.

The asymptotic mean square error of $\hat{h}\left(\underset{\sim}{z}\right)$ is finally derived as

$$AMSE\left[\hat{h}\left(\underset{\sim}{z}\right)\right] = n^{-1}d^{-v}\mathscr{R}(K)h\left(\underset{\sim}{z}\right) + \frac{1}{4}h^4[m_2(K)]^2 tr\left(D^2 h\left(\underset{\sim}{z}\right)\right) \tag{20}$$

after parametrization of **M** by taking $d_1 = \ldots = d_v = d(>2)$.

2.2 Kernel density estimation in a partially nonlinear model

By taking $\psi(X_1, \ldots, X_u; \beta_1, \ldots, \beta_u)$ a nonlinear function of the covariates $X_1, \ldots, X_u$, a partially nonlinear regression model is obtained. The error assumptions remain the same as in a partially linear model; however, the model estimation requires a slightly different technique other than Robinson's approach, which is solely applicable to a linear parametric regression function. In this setup, a traditional least squares estimation of parameters can be a convenient way, followed by kernel density estimation of $h(Z_1, \ldots, Z_v)$. Let us take the error sum of squares of the model under consideration as

$$S^2 = \sum_{i=1}^{n} [Y_i - \psi(X_{1i}, \ldots, X_{ui}; \beta_1, \ldots, \beta_u) - h(Z_{1i}, \ldots, Z_{vi})]^2 \tag{21}$$

and minimization of S^2 would yield the least squares estimators of $\beta_1, \ldots, \beta_u$ eventually, with the aid of various numerical methods. The procedure is elaborated next:

$$\frac{\delta S^2}{\delta \beta_j} = 0 \Rightarrow \sum_{i=1}^{n}(-2)\left[Y_i - \psi\left(\underset{\sim}{X_i}; \underset{\sim}{\beta}\right) - h\left(\underset{\sim}{Z_i}\right)\right] \times \psi_{\beta_j}\left(\underset{\sim}{X_i}; \underset{\sim}{\beta}\right) = 0; j = 1, \ldots, u \tag{22}$$

$$\Rightarrow \sum_{i=1}^{n} Y_i \psi_{\beta_j}\left(\underset{\sim}{X_i}; \underset{\sim}{\beta}\right) = \sum_{i=1}^{n} \psi\left(\underset{\sim}{X_i}; \underset{\sim}{\beta}\right)\psi_{\beta_j}\left(\underset{\sim}{X_i}; \underset{\sim}{\beta}\right) + \sum_{i=1}^{n} h\left(\underset{\sim}{Z_i}\right)\psi_{\beta_j}\left(\underset{\sim}{X_i}; \underset{\sim}{\beta}\right) \tag{23}$$

for all $j = 1, \ldots, u$, which are u normal equations to be solved for computation of the parameters. Here $\psi_{\beta_j}(\cdot, \ldots, \cdot)$ is the first order partial derivative of $\psi(\cdot, \ldots, \cdot)$ with respect to β_j for $j = 1, \ldots, u$.

After the estimated values of $\beta_1, \ldots, \beta_u$ are computed as $\hat{\beta}_1, \ldots, \hat{\beta}_u$, we proceed to transform the response Y to $Y' = Y - \sum_{i=1}^{n} \psi\left(\underset{\sim}{X_i}; \underset{\sim}{\hat{\beta}}\right)$, similar to the case of the partially linear model's transformed response Y' with only a difference in the nonlinearity of the parametric explanatory part of Y. Next, the usual Nadaraya-Watson kernel density estimation of $h(Z_1, \ldots, Z_v)$ is performed, yielding its Nadaraya-Watson kernel density estimator as

$$\hat{h}(z_1, \ldots, z_v) = \sum_{i=1}^{n} \omega_i^* Y_i' \tag{24}$$

where $\omega_i^* = \dfrac{\left\{\prod_{j=1}^{v} k_{Z_j}\left(\frac{z_j - Z_{ji}}{d_j}\right)\right\}}{\sum_{i=1}^{n}\left\{\prod_{j=1}^{v} k_{Z_j}\left(\frac{z_j - Z_{ji}}{d_j}\right)\right\}}, i = 1, \ldots, n.$

3. Local regression in semiparametric regression

3.1 Local linear smoothing of regression function

As usual, we estimate $\underset{\sim}{\beta} = (\beta_1, \dots, \beta_u)^T$ at the preliminary step by Robinson's technique, followed by minimization of the following local weighted sum of squares

$$D\left(a_1, \underset{\sim}{a_2}\right) = \sum_{i=1}^{n} \left(Y_i' - a_1 - \underset{\sim}{a_2^T}\left(\underset{\sim}{Z_i} - \underset{\sim}{z}\right)\right)^2 |H_Z|^{-1} K\left(|H_Z|^{-1}\left(\underset{\sim}{Z_i} - \underset{\sim}{z}\right)\right) \tag{25}$$

where the nonparametric regression function $h(Z_1, \dots, Z_v)$ is approximated as $h\left(\underset{\sim}{Z}\right) \approx a_1 + \underset{\sim}{a_2^T}\underset{\sim}{Z}, \underset{\sim}{Z} = (Z_1, \dots, Z_v)^T.K(\cdot)$ is the v-dimensional kernel density function and H_Z is the suitably formed non-singular bandwidth matrix of order v. $a_1, \underset{\sim}{a_2} \in \mathbb{R}^v$ are the parameters and $Y_i' = Y_i - \sum_{j=1}^{u} \hat{\beta}_j X_j, i = 1, \dots, n$. Also, $\underset{\sim}{Z_i} = (Z_{1i}, \dots, Z_{vi})^T$ is the ith random sample of $\underset{\sim}{Z}, 1 \le i \le n.\underset{\sim}{\beta_j}$ is the estimate of j-th component β_j in $\underset{\sim}{\beta}; j =$

$1, \dots, u$. Now, defining $\underset{\sim}{a_2} = (a_{21}, \dots, a_{2v})^T$, we minimize $D\left(a_1, \underset{\sim}{a_2}\right)$ with respect to $a_1, a_{21}, \dots, a_{2v}$ and later obtain the least squares estimates of a_{2p}'s; $p = 1, \dots, v$.

Note that,

$$\frac{\delta D}{\delta a_1} = 0 \Rightarrow \frac{\delta}{\delta a} \sum_{i=1}^{n} \left(Y_i' - a_1 - \sum_{s=1}^{v} a_{2s}(Z_{si} - z_s)\right)^2 \times W_{Z_i}\left(\underset{\sim}{z}\right) = 0 \tag{26}$$

$$\Rightarrow \sum_{i=1}^{n} (-2)\left[Y_i' - a_1 - \sum_{s=1}^{v} a_{2s}(Z_{si} - z_s)\right] \times W_{Z_i}\left(\underset{\sim}{z}\right) = 0 \tag{27}$$

$$\Rightarrow \sum_{i=1}^{n} Y_i' W_{Z_i}\left(\underset{\sim}{z}\right) = a_1 \sum_{i=1}^{n} W_{Z_i}\left(\underset{\sim}{z}\right) + \sum_{s=1}^{v} a_{2s}\left\{\sum_{i=1}^{n} (Z_{si} - z_s)\right\} W_{Z_i}\left(\underset{\sim}{z}\right) \tag{28}$$

and,

$$\frac{\delta D}{\delta a_{2t}} = 0 \Rightarrow \frac{\delta}{\delta a_{2t}} \sum_{i=1}^{n} \left(Y_i' - a_1 - \sum_{s=1}^{v} a_{2s}(Z_{si} - z_s)\right)^2 \times W_{Z_i}\left(\underset{\sim}{z}\right) = 0 \tag{29}$$

$$\Rightarrow \sum_{i=1}^{n} (-2)(Z_{ti} - z_t)\left[Y_i' - a_1 - \sum_{s=1}^{v} a_{2s}(Z_{si} - z_s)\right] \times W_{Z_i}\left(\underset{\sim}{z}\right) = 0 \tag{30}$$

$$\Rightarrow \sum_{i=1}^{n} (Z_{ti} - z_t) Y_i' W_{Z_i}\left(\underset{\sim}{z}\right) = a_1 \theta_t + \sum_{s=1}^{v} a_{2s}(Z_{ti} - z_t)\alpha_{ts} \tag{31}$$

where $\theta_t = \sum_{i=1}^{n} (Z_{ti} - z_t) W_{Z_i}\left(\underset{\sim}{z}\right), \alpha_{ts} = \sum_{i=1}^{n} (Z_{ti} - z_t)(Z_{si} - z_s) W_{Z_i}\left(\underset{\sim}{z}\right)$ and $W_{Z_i}\left(\underset{\sim}{z}\right) = |H_Z|^{-1} K\left(|H_Z|^{-1}\left(\underset{\sim}{Z_i} - \underset{\sim}{z}\right)\right)$ for $t, s = 1, \dots, v, i = 1, \dots, n$.

Furthermore, we shall solve the following $(v + 1)$ equations in the vector-matrix form

$$\underset{\sim}{Y}^* = M\underset{\sim}{a} \tag{32}$$

where

$$\underset{\sim}{Y}^* = \left(\sum_{i=1}^{n} Y_i' W_{Z_i}\left(\underset{\sim}{z}\right), \sum_{i=1}^{n}(Z_{1i} - z_1) Y_i' W_{Z_i}\left(\underset{\sim}{z}\right), \dots, \sum_{i=1}^{n}(Z_{vi} - z_v) Y_i' W_{Z_i}\left(\underset{\sim}{z}\right)\right)^T, \underset{\sim}{a} = (a_1, a_{21}, \dots, a_{2v})^T$$

and

$$M^{v+1} = \begin{pmatrix} n & \theta_1 & \dots & \theta_v \\ \theta_1 & \alpha_{11} & \dots & \alpha_{1v} \\ \vdots & \vdots & & \vdots \\ \theta_v & \alpha_{v1} & \dots & \alpha_{vv} \end{pmatrix}$$

and the solution of (32) is obtained as

$$\hat{a}_{\sim GLS} = M^{-}\underset{\sim}{Y}^* \tag{33}$$

where M^- denotes the generalized inverse of M provided that $rank(M) < v + 1$, and $\hat{a}_{\sim GLS}$ is the generalized least squares estimate of $\underset{\sim}{a}$. When M is non-singular, that is, $rank(M) = (v + 1)$, then

$$\hat{a}_{\sim LS} = M^{-1}\underset{\sim}{Y}^* \tag{34}$$

where M^{-1} is the inverse of M. $\hat{a}_{\sim LS}$ is the least squares estimate of $\underset{\sim}{a}$.

3.2 Local polynomial smoothing of regression function

In this context, the nonparametric regression function $h(Z_1, \dots, Z_v)$ is approximated by a polynomial of suitable degree, the one for which the residual sum of squares $RSS = \sum_{i=1}^{n}\left(Y_i' - h\left(\underset{\sim}{Z_i}\right)\right)^2$ is minimum with $Y' = Y - \sum_{j=1}^{u}\hat{\beta}_j X_j, j = 1, \dots, u$.

Proceeding with the following assumption of $h\left(\underset{\sim}{Z_i}\right)$ for $i = , \dots, n$ as

$$h(Z_{1i}, \dots, Z_{vi}) \approx a_0 + \{a_{11}(z_1 - Z_{1i}) + \dots + a_{1v}(z_v - Z_{vi})\} \tag{35}$$

$$+\left\{a_{21}(z_1 - Z_{1i})^2 + \dots + a_{2v}(z_v - Z_{vi})^2\right\} \tag{36}$$

$$+ \dots + \{a_{p1}(z_1 - Z_{1i})^p + \dots + a_{pv}(z_v - Z_{vi})^p\} \tag{37}$$

the following weighted sum of squares needs to be minimized with respect to $a_0, a_{11}, \dots, a_{1v}, \dots, a_{p1}, \dots, a_{pv}$ as

$$D\left(\underset{\sim}{a_0}, a_1, \dots, \underset{\sim}{a_p}\right) = \sum_{i=1}^{n} \left(Y_i' - h(Z_{1i}, \dots, Z_{vi})\right)^2 \tag{38}$$

$$\frac{\delta D}{\delta a_0} = 0 \Rightarrow \sum_{i=1}^{n} Y_i' = na_0 + \sum_{i=1}^{n}\sum_{j=1}^{v}\sum_{r=1}^{p} a_{rj}\left(z_j - Z_{ji}\right)^r \tag{39}$$

and

$$\frac{\delta D}{\delta a_{r'j'}} = 0 \Rightarrow \sum_{i=1}^{n} Y_i'\left(z_{j'} - Z_{j'i}\right)^{r'} = a_0 \sum_{i=1}^{n}\left(z_{j'} - Z_{j'i}\right)^{r'} + \left(z_{j'} - Z_{j'i}\right)^{r'} \sum_{i=1}^{n}\sum_{j=1}^{v}\sum_{r=1}^{p} a_{rj}\left(z_j - Z_{ji}\right)^r \tag{40}$$

for all $j' = 1, \dots, v, r' = 1, \dots, p$.

By solving the above equations, the estimated values of $\underset{\sim}{a_0}, a_1, \dots, \underset{\sim}{a_p}$ are obtained as $\underset{\sim}{\hat{a}_0}, \hat{a}_1, \dots, \underset{\sim}{\hat{a}_p}$ and the polynomial regression function $h(\cdot)$ is estimated after plugging in the values of the estimated values of the parameters as

$$\hat{h}(Z_{1i}, \dots, Z_{vi}) = \hat{a}_0 + \{\hat{a}_{11}(z_1 - Z_{1i}) + \dots + \hat{a}_{1v}(z_v - Z_{vi})\} \tag{41}$$

$$+\left\{\hat{a}_{21}(z_1 - Z_{1i})^2 + \dots + \hat{a}_{2v}(z_v - Z_{vi})^2\right\} \tag{42}$$

$$+ \dots + \left\{\hat{a}_{p1}(z_1 - Z_{1i})^p + \dots + \hat{a}_{pv}(z_v - Z_{vi})^p\right\}. \tag{43}$$

4. Spline smoothing in semiparametric regression

In the transformed regression setup $Y' = h(Z_1, \dots, Z_v) + \epsilon$ one may apply the spline smoothing technique to obtain a reasonable estimator of $h\left(\underset{\sim}{Z}\right)$. Generally, as we have discussed in this discourse that minimization of residual sum of squares $\sum_{i=1}^{n}\left(Y_i' - h\left(\underset{\sim}{Z_i}\right)\right)^2$ would yield a meaningful estimate of $h(\cdot)$ as $\hat{h}(\cdot)$, that is not a unique estimator fo $h\left(\underset{\sim}{Z_i}\right)$ for all $i = 1, \dots, n$. In fact, one may be able to derive a curve $h^*\left(\underset{\sim}{Z_i}\right)$ such that $Y_i' - h^*\left(\underset{\sim}{Z_i}\right) = 0$ for all $i = 1, \dots, n$; hence, the usual sum of squares minimization technique does not serve the purpose of estimating $h(\cdot)$ in a proper, unambiguous manner. Instead of considering all the datapoints on $Z_1, \dots, Z_v$, it is more rational to control the local variation of the curve $h(\cdot)$ on every interval $\left(Z_{ks}, Z_{k(s+1)}\right) \subseteq \left[\min_{1 \le i \le n} Z_{ki}, \min_{1 \le i \le n} Z_{ki}\right]$ for $1 \le s \le n-1, k = 1, \dots, v$. To quantify the local variations of $h(\cdot)$, the roughness penalty (also known as *integrated squared Laplacian* due to Friedman (1991)) $\mathscr{R}(h) = \sum_{i=1}^{n}\sum_{j=1}^{n} \int_{\mathbb{R}^q} D_{ij}^2 h \, d\underset{\sim}{z}$ is utilized where $D_{ij}^2 \equiv \frac{\delta^2}{\delta z_i \delta z_j}; i, j = 1, \dots, n$.

It is noteworthy that the appropriate smoothing curve for $h\left(\underset{\sim}{Z}\right)$ can be determined by minimizing the following weighted sum of squares

$$S_\lambda(h) = \sum_{i=1}^{n} \left[Y_i' - h\left(\underset{\sim}{Z_i}\right) \right]^2 + \lambda \mathscr{R}(h) \tag{44}$$

where λ is the parameter delineating the trade-off between *lack of fit of data* and *roughness of* $h(\cdot)$. The estimated curve $h_\lambda(\cdot)$ is determined as

$$h_\lambda\left(\underset{\sim}{Z_i}\right) = arg \min_h S_\lambda(h) \tag{45}$$

$$= arg \min_h \sum_{i=1}^{n} \left[Y_i' - h\left(\underset{\sim}{Z_i}\right) \right]^2 + \lambda \mathscr{R}(h); i = 1, \dots, n \tag{46}$$

and in $\left(Z_{ks}, Z_{k(s+1)}\right)$ $(s = 1, \dots, (n-1); k \in \{1, \dots, v\})$ the curve $h_\lambda(\cdot)$ is a polynomial of degree 3, that is,

$$h_\lambda\left(\underset{\sim}{Z_i}\right) = a_0 + \sum_{r=1}^{3} \sum_{s=1}^{v} a_{rs} (z_s - Z_{si})^r ; i = 1, \dots, n. \tag{47}$$

5. Tables

See **Tables 1–3**.

X_1	X_2	Z_1	Z_2	$\hat{h}\left(\underset{\sim}{Z_1, Z_2}\right)$	$\hat{Y}'$
−1.140	0.882	−1.735	−0.428	−2.128	−1.743
−0.733	1.776	−1.731	0.337	−0.027	0.820
−0.553	0.692	−1.629	−0.945	0.257	0.575
−0.651	0.749	−1.538	−0.683	−3.362	−3.024
−1.021	1.469	−1.365	0.674	−0.368	0.313
−0.117	1.130	−1.361	−0.388	0.982	1.539
$\hat{\beta}_1$				$\hat{\beta}_2$	
0.047				0.497	

Table 1.
Kernel density estimation of $h(Z_1, Z_2)$ *using Gaussian kernel and estimation of* β_1, β_2 *in a generalized partially linear model* $Y = \beta_1 X_1 + \beta_2 X_2 + h(Z_1, Z_2) + \epsilon$.

X_1	X_2	Z_1	Z_2	$\hat{h}\left(\underset{\sim}{Z_1, Z_2}\right)$	$\hat{Y}'$
−1.315	1.468	−1.877	0.205	0.502	0.701
−0.761	1.109	−1.761	−0.472	1.155	1.217
−1.131	0.351	−1.689	−1.004	−0.639	−0.282
0.104	0.512	−1.631	−1.645	0.042	−0.103
−0.576	1.312	−1.565	−0.107	−0.589	−0.639

X_1	X_2	Z_1	Z_2	$\hat{h}\left(\underset{\sim}{Z_1, Z_2}\right)$	$\hat{Y}'$
−1.215	0.639	−1.523	−0.378	0.261	0.591
$\hat{\beta}_1$				$\hat{\beta}_2$	
−0.379				−0.204	

Table 2.
Local polynomial (of degree 2) smoothing estimation of $h(Z_1, Z_2)$ *using Gaussian kernel and estimation of* β_1, β_2 *in a generalized partially linear model* $Y = \beta_1 X_1 + \beta_2 X_2 + h(Z_1, Z_2) + \epsilon$.

X_1	X_2	Z_1	Z_2	$\hat{h}\left(\underset{\sim}{Z_1, Z_2}\right)$	$\hat{Y}'$
−1.479	0.693	−1.616	−0.244	2.563	3.152
−0.599	0.042	−1.561	−1.591	−1.559	−1.272
−0.654	0.838	−1.511	−0.540	0.387	0.540
0.019	0.404	−1.506	−1.544	−1.957	−2.048
−1.030	0.219	−1.482	−0.958	0.026	0.489
−1.082	1.319	−1.474	0.395	0.134	0.402
−0.168	0.573	−1.437	−1.113	1.158	1.126
$\hat{\beta}_1$				$\hat{\beta}_2$	
−0.492				−0.301	

Table 3.
Spline smoothing estimation of $h(Z_1, Z_2)$ *using Gaussian kernel and estimation of* β_1, β_2 *in a generalized partially linear model* $Y = \beta_1 X_1 + \beta_2 X_2 + h(Z_1, Z_2) + \epsilon$.

6. Conclusions

Throughout this chapter, we have proposed some suitable estimation techniques for the parametric and nonparametric components of a partially linear model. Various nonparametric methods have been used in this chapter, *viz.* kernel density estimation, spline smoothing estimation, local polynomial regression, etc. The estimation procedures are delineated elaborately in the sections 1–3. The estimated values of the regression function under a generalized partially linear model $Y = \beta_1 X_1 + \beta_2 X_2 + h(Z_1, Z_2) + \epsilon$ are obtained in Section 5. In general, the kernel smoothing technique is preferred over other methods. Furthermore, one can compute the asymptotic mean square error of the estimated nonparametric regression function. All the semiparametric estimators of the model components are consistent as well as they are asymptotically Gaussian [2, 15].

Author details

Sthitadhi Das
Brainware University, Kolkata, India

*Address all correspondence to: sthdas999@gmail.com

References

[1] Speckman P. Kernel smoothing in partial linear models. Journal of the Royal Statistical Society Series B: Statistical Methodology. 1988;**50**(3): 413-436

[2] Li Q. On the root-n-consistent semiparametric estimation of partially linear models. Economics Letters. 1996; **51**(3):277-285

[3] Robinson PM. Root-N-consistent semiparametric regression. Econometrica: Journal of the Econometric Society. 1988;**56**:931-954

[4] Powell JL. Semiparametric estimation. In: Microeconometrics. London, UK: Palgrave Macmillan; 1989. pp. 267-277

[5] Pfanzagl J, Pfanzagl J. Estimation in Semiparametric Models. Vol. 60. US: Springer; 1990. pp. 17-22

[6] Härdle W, Müller M, Sperlich S, Werwatz A. Nonparametric and Semiparametric Models. Vol. 1. Berlin: Springer; 2004

[7] Bickel PJ, Klaassen CA, Bickel PJ, Ritov YA, Klaassen J, Wellner JA, et al. Efficient and Adaptive Estimation for Semiparametric Models. Vol. 4. Baltimore: Johns Hopkins University Press; 1993

[8] Ruppert D. Semiparametric Regression. Cambridge: Cambridge University Press; 2003

[9] Lee S. Efficient semiparametric estimation of a partially linear quantile regression model. Econometric Theory. 2003;**19**(1):1-31

[10] Liang H. Estimation in partially linear models and numerical comparisons. Computational Statistics & Data Analysis. 2006;**50**(3):675-687

[11] Li Q. Efficient estimation of additive partially linear models. International Economic Review. 2000;**41**(4): 1073-1092

[12] Schimek MG. Estimation and inference in partially linear models with smoothing splines. Journal of Statistical Planning and Inference. 2000;**91**(2): 525-540

[13] Hamilton SA, Truong YK. Local linear estimation in partly linear models. Journal of Multivariate Analysis. 1997; **60**(1):1-19

[14] Andrews DW. Nonparametric kernel estimation for semiparametric models. Econometric Theory. 1995;**11**(3): 560-586

[15] Das S, Maiti SI. On the test of association between nonparametric covariate and error in semiparametric regression model. Journal of the Indian Society for Probability and Statistics. 2022;**23**(2):541-564

Chapter 3

Statistical Regularities in the Musical Work of Marin Marais, Pièces de Viole Des Cinq Livres

Igor Lugo and Martha G. Alatriste-Contreras

Abstract

This study analyzes the spectrum of audio signals related to the work of "Pièces de viole des Cinq Livres" based on the music of Marin Marais, performed by Jordi Savall. In particular, we aim to identify the statistical regularity underlying this musical work. Based on the complex systems approach, we compute the spectrum of audio signals and analyze and identify their best-fit statistical distributions. Findings suggest that the collection of frequency components related to the spectrum of each of the audio-books shows highly skewed and associated statistical distributions, in particular the presence of the exponential statistical distribution.

Keywords: statistical regularities, empirical distributions, music, audio signal processing, viola da Gamba

1. Introduction

Marin Marais is one of the most outstanding composers and performer musicians of bass viola da gamba in music history. Not only in his era, but also in the current days, his music has continued to delight from novice to expert musicians. Nowadays, Jordi Savall—a contemporary conductor, composer, historian, and viol player—has rescued most of Marias' work from oblivion and communicated it globally. The extraordinary musical contribution of both musicians has shown one of the highest levels of musical expression over time. There is a deep musical connection between this pair of musicians (named in the following as Marais-Savall) that has transcended time and frontiers. However, a scientific analysis of Marais-Savall audio signals—waveforms—is still missing in the music and scientific communities. In particular, one of the most important works of Marais, "Pièces de viole des Cinq Livres," has not been explored based on its statistical properties underlying its music information. Therefore, our chapter aims to analyze the audio signals of this musical work for identifying the statistical distributions that best describe their spectrum—the frequency of components related to musical notes. After establishing the presence of such distributions, we can reinterpret the music of bass viol and identify the statistical regularities of viol players.

This statistical approach for identifying consistent and predictable patterns in music has been applied by Lugo and Alatriste-Contreras [1]. They suggested that the concept of virtuosity in music is possibly related to entropy values and the best-fit distributions of the spectrum of audio signals. In particular, they suggested that the waveform and its spectrum contain information to identify levels of virtuosity in music. Moreover, the work of Downey [2] showed different topics of signal processing in music. In particular, he presented techniques and applications with a programming-based approach for understanding real audio signals. Other relevant work of audio signals is Müller and Klapuri [3]. They presented an overview of principles and applications of music signals that are the key for underlying music analysis problems. Therefore, the data analysis of audio signals based on interdisciplinary approaches and the current digital technology may generate a deep understanding of the underlying information in music. The intuition of musicians about identifying a particular composer when only playing or hearing a few notes of some music repertoire can be confirmed if we look into the statistics of the spectrum of signals.

In the case of Marais-Savall, the identification of their unique statistical pattern or signature might be related to different aspects of their lives. Regarding Marais, several studies on his musical abilities have highlighted factors that are associated with his personal experience and social relationships [4, 5]. In particular, the work of Milliot and de la Gorce [6] described almost a complete view of the context in which Marais developed his musical creativity and skills. For example, his relationship with two of the most respected musicians of that time, Jean de Sainte-Colombe and Jean-Baptiste Lully. Other work that complemented this reference is the audio work of Savall et al. [7]. It offers materials that can not only be listened to but also read; we can read information about the collection of the audio tracks. For example, the booklet described the common discussion among musicians at that time about the balance between melody and harmony. Finally, an unexpected yet interesting study is the work of Matloubieh et al. [8]. This study came from another discipline, and the authors suggested that Marais mixed his musical reputation with a medical procedure about lithotomy. Then, the influence of Marais covered not only common issues in music but also different and relevant themes of his time.

In the case of Savall, there are different information sources that display his work across several areas—that is, concert performer, teacher, researcher, just to name a few [9]. For example, the websites of AliaVox and Fundació Centre International de Música Antiga illustrate his outstanding works in rescuing and preserving early music. The work of Forti i Murrugat [10] analyzed Savall's projects related to this type of music to propose a musical framework for music and art. Therefore, the Marais-Savall relationship shows singular musical and personal characteristics that are possibly imprinted in most of their musical work.

Our main questions are the following: Does the collection of the five books show similar statistical distributions? and, what are those statistical distributions? We believe that the underlying attributes of waveforms and their audio spectrum are related to statistical distributions. They can be interpreted as the signature of a set of music repertoire. Depending on the performance of musicians—different musicians play the same repertoire—the signature varies only slightly from its real value. Therefore, a large set of audio signals of bass viol played by the same musician provides the event-based condition for identifying accurately the signature of a particular musical work.

The document is divided into four sections. The materials section shows the audio resources from the data that was collected. The methods section explains the application of the complex systems approach to an explorative data analysis based on identifying statistical distributions. The results section displays our findings. Finally, the discussion section points out some items to be considered in the analysis and gives the conclusions.

2. Material

The data consisted of the audio material related to the work of Savall et al. [7]. This material, named Pièces de viole des Cinq Livres, is a collection of five audio CDs, containing a total of 84 tracks. We selected this audio material because it is one of the best recording audio data about Marin Marais up until now. This material combines together audio recordings and historical documents that more accurately describe the Marais' musical contribution. As a listener, the execution behind each track reflects Savall's expertise in playing the instrument and his knowledge of recording music. Because of the copyrights of this audio material, we suggest obtaining these CDs and following our method for replicating results. Therefore, we transformed each track of this album from *m4a* to *WAV* files in 16-bit PCM. During this process, we spliced stereo to mono using Audacity.

On the other hand, we used different Python libraries to retrieve, analyze, and plot the data and results. In particular, we used NUMPY, MATPLOTLIB, SCIPY, and PANDAS. Moreover, we used some parts of the code provided by Downey [2]. The database and the code are available in our Open Science Framework (OSF) for the reproduction of our findings: Complex systems and early music, DOI 10.17605/OSF.IO/2BYQV.

3. Methods

High-quality audio recordings of viola da gamba are rare because it is not common to play such an instrument nowadays. In this case, Savall has provided a unique collection of recordings of Marais that we can analyze based on their audio signals. Therefore, in this section, we present our procedure for identifying the statistical signatures of the collaborative work between Marais and Savall.

The first step in this procedure is the use of the spectral decomposition. This is a procedure for simplifying audio data based on the Fast Fourier Transformation (FFT) algorithm and the Discrete Fourier Transformation (DFT) [11–13]. The result of this decomposition is named the "spectrum," which shows approximately the frequency components related to pitches—the dominant pitch and its harmonics. The importance of this spectrum is to identify the frequency components of sections related to the immediate musical composition or improvisation. The analysis of these sections is not trivial because they are commonly related to different musical interpretations in which the musician communicates emotions through their performance. For example, contemporary guitarists improvise solos that engage audiences or listeners to experience different emotions [14]. Then, the spectrum is one of the keys to unfolding the rich musical imaginations of celebrated composer-performer musicians [15]. Therefore, we used the spectral decomposition associated with the FFT as follows:

$$y[k] = \sum_{n=0}^{N-1} e^{-2\pi j \frac{kn}{N}} x[n] \qquad (1)$$

where $y[k]$ is the frequency component of a sequence of the signal $x[n]$ from n to $N - 1$. In our case, the identification of these components provides the inputs for understanding the statistical signature associated with our audio material. Therefore, we can define the statistical signature of a collection of audio materials as a clearly identified statistical distribution that shows particular properties.

Next, the second step is to analyze those collections of frequency components looking for the identification of statistical distributions that best describe them. The statistical distribution or probability distribution is a mathematical function that describes the occurrence of possible events. It approximates the generating process of particular data. A major advantage of this function is to infer properties underlying it [16, 17]. In particular, continuous distributions are commonly related to three types of parameters: location, scale, and shape [18]. The location parameter, which is associated with the first moment or mean, refers to the place where the most frequent value is observed along the x-axis in a frequency plot. The scale parameter, which is associated with the second moment or variance, refers to how spread out are the data with respect to the location along the x-axis. The shape parameter, which is associated with all higher moments such as the skewness and kurtosis, refers to the shape or geometry of the data. It is important to mention that depending on the skewed or non-skewed data, the location and scale parameters can be related to different measures; for example, for skewed data, it is commonly suggested to compute the median and the entropy [19–21]. Therefore, based on this collection of statistical measures, we can describe and infer the behavior of any distribution. In our case, after obtaining the frequency components, we are in the possibility of identifying the statistical distribution that represents accurately the unique audio signal of the work of Marais-Savall.

To identify the possible statistical distributions that best describe the audio work of Marais-Savall, we used the Kolmogorov-Smirnov (KS) test to identify whether or not our audio data comes from a certain distribution [22]. In our case, this test compares our frequency components (empirical data) with a set of given distributions (theoretical distributions). We used the normal, log-normal, exponential, Pareto, Gilbrat, power law, and exponentiated Weibull as our theoretical distributions. These distributions represent an important set of continuous distributions in the literature that might represent non-skewed and skewed data, as well as different relationships between them [23, 24]. In particular, these distributions are connected with other distributions based on their properties, for example, the linear combination, coevolution, and products, just to mention a few [18]. Therefore, the KS test using those statistical distributions provides an accurate process to compute and identify distributions that best describe the audio work of Marais-Savall.

Finally, once the best-fit distributions were identified, we display their relative frequencies of the frequency components using the scientific pitch notation—note names and octave numbers—as bins (**Figure 1**). This figure aims to show the statistical attributes of audio signals based on how frequently some notes and octaves are used. It is the key to understand the possible statistical signature of the work of Marais-Savall. In the results section, we display this figure in each result related to the book's audio data.

In essence, we use each track per book for computing the FFT. The resulting book's collection of component frequencies is analyzed for identifying its best-fit statistical distribution. To identify the distribution that best describes the component frequencies, we show our proposed plot of relative frequencies of the spectrum (**Figure 1**).

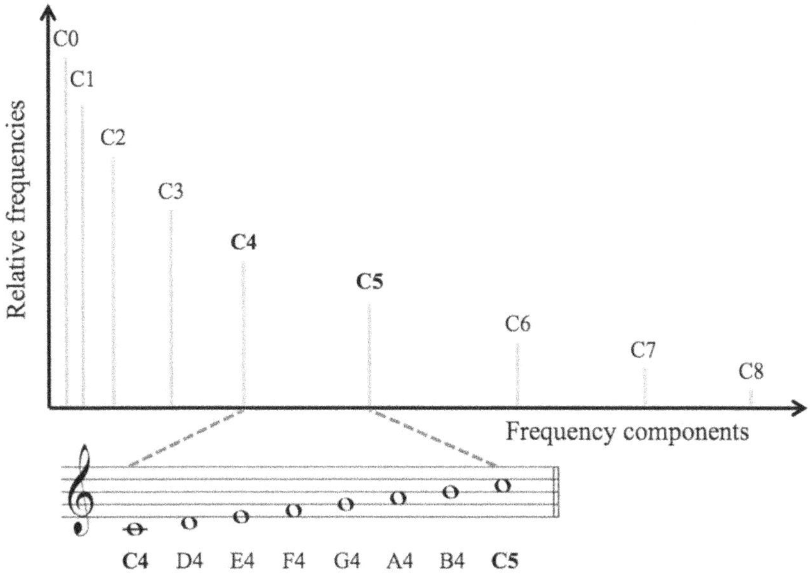

Figure 1.
Relative frequencies of the spectrum. Relative frequencies are related to Hz, and frequency components are associated with the output of the FFT (Eq. 2). We used the data provided by the Physics Department, at Michigan Technological University [25]. To avoid confusion in the scientific pitch notation, notes are translated by multiplying or dividing the frequency by 2. Then, in this figure, vertical lines are approximations of octave locations used for reference purposes only.

4. Results

As we mentioned previously, the main goal of this chapter is to identify possible statistical regularities related to the spectrum of audio signals of the work of "Pièces de viole des Cinq Livres" based on the work of Marais-Savall. Each book, which is a collection of audio tracks, of this work was analyzed by following our proposed method. Therefore, for simplicity and ease of interpretation, we are going to show our results for the five books by using only our proposed plot related to the relative frequencies (**Figure 1**).

Figure 2 shows the statistical signature of each of the books. As can be seen, the curves related to each of the collection of frequency components associated with their spectrums are highly skewed. This lack of symmetry suggested that lower and higher octaves are played more frequently. In particular, lower octaves represent the majority of notes played between the $C0$ and $C3$. On the other hand, higher octaves represent the relative notes played between $C6$ and $C8$. Moreover, between $C3$ and $C6$ octaves, we can see important differences in played notes. Books 4 and 5 represent the greatest difference; meanwhile, books 1, 2, and 3 are in between them. This particular result suggests that the main differences between the frequency components of each book are around the standard tuning $C4$.

In addition to these results, we present the estimated parameters related to the location, scale, and shape per book. As we can see in **Table 1**, there are two values related to the location: median and mean. In this particular case, we are interested in the median due to the resulting highly skewed distributions. Using the mean value can be misleading or incorrect because it is commonly related to non-skewed data, such as the normal distribution. Then, the median values of the books associated with the exponential distribution show a stable location between the notes $A1$ and $E2$;

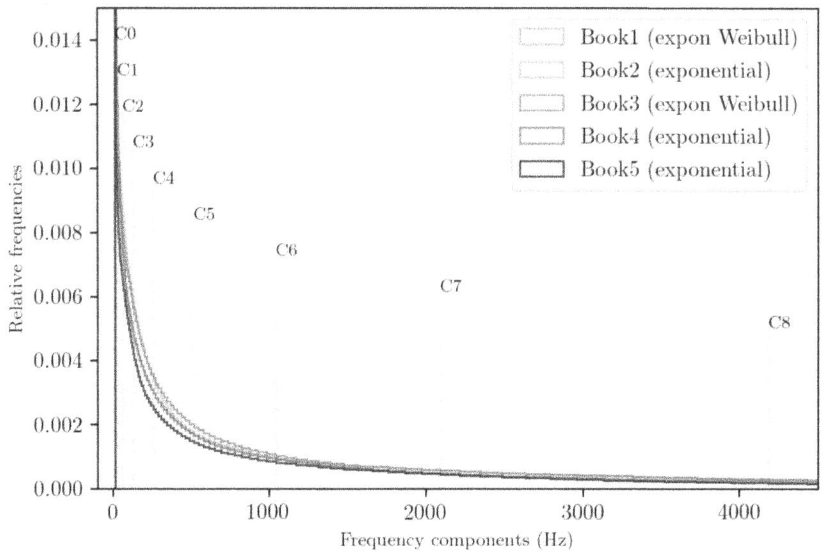

Figure 2.
*Relative frequencies of the audio spectrums and best-fit distributions. See **Table** 2 for statistical results of best-fit, parameters, and KS test results.*

Name	Median	Mean	Variance	Entropy	Skew	Kurtosis
Book 1	11.9226	15.9130	165.1470	2.2294	8.6444	165.0793
Book 2	79.3857	114.5291	13116.7683	5.7408	2.0	6.0
Book 3	19.4983	58.5914	15103.0876	4.7947	7.5989	126.6237
Book 4	77.2398	111.4334	12417.3580	5.7134	2.0	6.0
Book 5	54.4538	78.5601	6171.6320	5.3638	2.0	6.0

Table 1.
Estimated parameters of frequency components per book.

meanwhile, *book* 1 and *book* 3 showed locations less than C0. Next, there are two scale parameters related to the data dispersion: variance and entropy. As we have just mentioned, we used the entropy value due to its attributes related to skewed data. Then, entropy values showed similar dispersion except for *book* 1. Finally, the estimated shape parameters related to geometry showed that there is more weight on the right tail of the distributions, and they exhibit peaked shapes.

Summing up, the results of our data analysis are conclusive for identifying reliable parameters that point out the statistical signature of the work of Marais-Savall. The frequency components of the audio spectrums related to each book were associated with highly skewed distributions. These distributions may well be related to the exponential distribution because it is the most frequent best-fit distribution presented in our results.

5. Discussion

Our study about the musical collaboration between Marais and Savall related to the audio work of "Pièces de viole des Cinq Livres" has shown the possibility of underlying its musical information. In particular, we could identify statistical attributes that distinguish the most frequent best-fit distributions related to each book's audio spectrums. Consequently, our results indicated that the frequency components of such spectrums must have been related to the presence of highly skewed distributions, particularly in relationship with the exponential statistical distribution.

The significance of these findings is to be found in recognition of the musical work between musicians. Even though musicians are separated by time periods and places, their original and unique musical contributions can be recognized not only by the timbre, but also by the information related to the audio wave. In our case, the collaboration between Marin Marais and Jordi Savall has shown one of the highest levels of musical expressions over time that must have been recognized by their highly skewed distributions of the frequency components of audio spectrums. Such statistical distributions are related to the exponential distribution. This type of distribution is commonly used to describe system reliability and the times between events [23]. One of its main characteristics is a constant failure rate function—no memory when considering events based on its age. In our case, the memoryless or Markovian [26] property indicates that if the most frequent octave notes are played for s units of times, the probability that higher octave notes will play in additional time units is independent of s. In other words, the probability of playing lower or higher octave notes in an audio track is independent. Therefore, these findings suggest that the transition from one note to another in an audio track follows a random process based on the exponential distribution.

Translating this result into musical expressions, we can say that there may be a link between the improvisation and the selection of notes in a particular musical passage. In the case of Marais-Savall, we know that the composition and performance abilities of Marais were frequently associated with improvisation [7, 27]. Consequently, it is expected that Savall's interpretation and performance would reflect Marais' habits. These findings may help us to understand that the free performance of the musician may follow different random processes that most of the time are related to skewed statistical distributions.

The implications of these findings regarding the teaching and learning of music are related to composing and playing activities that cover not only the bass viola da gamba, but also any type of string instruments. For example, in a musical composition, musicians may use the information related to the type of statistical distribution to explore alternatives or extensions to conceive a piece of music. Depending on the type of skewed distribution, musicians must more frequently use higher and lower octaves to achieve most of their musical material. Therefore, before starting the process of composition, it is recommended to analyze previous personal works and the work of other musicians for obtaining a unique and original material. In the case of performing, the prior information about statistical distributions can provide the keys for improvising music in different styles and contexts. For example, if we know that the work of Marais-Savall is best described by an exponential distribution, we must play the patterns suggested by such a distribution. We must play more frequently higher and lower octaves; meanwhile, around the standard tuning, we can play notes for connecting and transitioning those octaves. Following this information, we can replicate the work of those musicians or generate our personal material.

Future studies on the current topic are therefore recommended. In particular, a natural extension of our findings is to explore the following questions: Is it possible to find the presence of similar best-fit distributions whether a particular musical passage is played by different musicians? How different or similar can be the statistical properties of each performance? On the other hand, a future study with more focus on a computational approximation for composing music based on our findings is therefore suggested. To answer these questions in future work, we suggest following the line of complex systems and music.

6. Conclusions

The greatest contribution of this study is to underlie statistical properties related to early music, composed and played by two outstanding musicians. Our method can be used for analyzing not only the bass viola da gamba but also other string instruments and musicians. The audio work of "Pièces de viole des Cinq Livres" showed highly skewed distributions possibly related to the exponential distribution. This type of statistical distribution may contain the keys to understand the elements of musical composition and performance of the bass viola da gamba.

Conflict of interest

The authors declare no conflict of interest.

Additional information

This chapter is related to the following preprint: Igor Lugo and Martha G. Alatriste-Contreras, Pièces de viole des Cinq Livres and their statistical signatures: the musical work of Marin Marais and Jordi Savall, arXiv:2404.18355. https://doi.org/10.48550/arXiv.2404.18355

Appendix A

See **Tables 2** and **3**. The criteria for selecting the final result in **Table 2** were as follows:

1. Errors in the estimation method. Based on the scipy function, scipy.stats.rv_continuous.fit, we used the Maximum Likelihood Estimation (MLE). If there were no errors, we selected the best fit; if there were errors, we selected the second best fit.

2. Visualizing the Cumulative Distribution Function (CDF). If the estimated values of the KS test (d, p-value) of the first and second best fit were the same, we plotted the empirical and theoretical CDFs. This ensures to identify the best-fit statistical distribution related to data.

Consequently, we found that the fit of Pareto estimations showed RuntimeErrors. Then, we had to visualize the CDFs for selecting the final results in **Table 2.** For greater precision of the KS test estimated parameters and to plot the CDFs, see and execute the code in Complex systems and early music. For the use of the same criteria applied to different scientific studies, see the works of Lugo I et al. [28–30].

Name	Best and second best fit	Parameters (a, b, loc, scale)	KS test (d, p-values)
Book 1	Pareto	(0.5000772790696841, −2.5313693644133393, 2.5320219554920884)	(8.810404621462098e-05, 0.7663125997309213)
	Exponentiated Weibull*	(1.8078554913188745, 0.40464609484912933, 10.52751466016218)	(8.810404621462098e-05, 0.7663125997309213)
Book 2	Exponential*	(0.0007007909083517199, 114.52846083233646)	(8.531262287569952e-05, 0.785822040345603)
	Pareto	(0.45694976715045765, −2.3484869507168717, 2.349187741621745)	(8.531262287581054e-05, 0.7858220403442739)
Book 3	Exponentiated Weibull*	(2.388182477333179, 0.40846064327782183, 0.0006176602639155107, 8.883106480102926)	(7.587900003558357e-05, 0.7629043858598745)
	Exponential	(0.0006176602639155108, 92.6733864816965)	(7.587900003563908e-05, 0.76290438585909)
Book 4	Exponential*	(0.00024278816105826503, 111.43319977150223)	(9.143090021945799e-05, 0.7129222750411135)
	Pareto	(0.4717388601895959, −2.384176737481692, 2.3844195256376968)	(9.143090021945799e-05, 0.7129222750411135)
Book 5	Exponential*	(0.00036681097553665327, 78.55973558262349)	(7.877884394258405e-05, 0.8129868745072603)
	Pareto	(0.4239939171603715, −1.6836235103637094, 1.683990321337741)	(7.877884394258405e-05, 0.8129868745072603)

**Statistics of the selected best-fit test.*

Table 2.
Statistical attributes of the "Pièces de viole des Cinq Livres."

Name	PDF
Exponential	$f(x) = exp(-x)$, for $x >= 0$
Pareto	$f(x, b) = \frac{b}{x^{b+1}}$, for $x >= 1, b > 0$
Exponentiated Weibull	$f(x, a, c) = ac[1 - exp(-x^c)]^{a-1} exp(-x^c)x^{c-1}$, for $x > 0, a > 0, c > 0$

Table 3.
Name of the statistical distributions and their PDF.

Name of the books, the best and second best-fit statistical distributions, the estimated parameters of the KS goodness-of-fit test, and the KS test two-sided statistic. See **Table 3** for the name of statistical distributions and their probability density function (PDF).

Author details

Igor Lugo[1*†] and Martha G. Alatriste-Contreras[2†]

1 CRIM, Universidad Nacional Autónoma de México, México

2 Facultad de Economía, Universidad Nacional Autónoma de México, Ciudad de México, México

*Address all correspondence to: igorlugo@crim.unam.mx

† These authors contributed equally.

References

[1] Lugo I, Alatriste-Contreras MG. An experiment with electric guitar signals for exploring the virtuosity based on the entropy of music. arXiv. 2024:1-28. Available from: https://arxiv.org/abs/2404.16259 [Accessed: November 12, 2024]

[2] Downey AB. Think DSP: Digital Signal Processing in Python. Needham, Massachusetts: Shroff Publishers & Distributors Pvt. Ltd; 2016

[3] Müller M, Klapuri A. Music signal processing. In: Trussell J, Srivastava A, Roy-Chowdhury AK, Srivastava A, Naylor PA, Chellappa R, Theodoridis S, editors. Academic Press Library in Signal Processing: Volume 4. Kidlington, Oxford UK: Elsevier; 2014. pp. 713-756. DOI: 10.1016/B978-0-12-396501-1.00027-3

[4] Cyr M. Marin Marais, the "basse continue" and a 'different manner' of composing for the viol. The Musical Times. 2016;**157**:49-61. Available from: https://www.jstor.org/stable/44862536

[5] Bane MA. Marin Marais and his public. Journal of the Viola da Gamba Society of America. 2018;**50**:24-48. Available from: https://www.vdgsa.org/_files/ugd/d493e7_4f60de21c70f422c9e5ab0d2c5aaced0.pdf

[6] Milliot S, de la Gorce J. Marin Marais. Paris, France: Fayard; 1991

[7] Savall J, Koopman T, Smith H, Coin C. Marin Marais, Pièces de viole des Cinq Livres. CD. Orne, France: Alia-Vox; 2010. Available from: https://www.alia-vox.com/en/catalogue/marin-marais-pieces-\de-viole-des-cinq-livres/

[8] Matloubieh JE, Eghbali M, Rabinowitz R. Blood gushing and musical screaming in Marin Marais' cystolithotomy. Urology. 2016;**141**: 60-63. Available from: https://www.goldjournal.net/article/S0090-4295(20)30372-1/fulltext

[9] Fernández T, Tamaro E. Biografia de Jordi Savall [Internet]. Barcelona, España: Biografias y Vidas; 2022. Available from: https://www.biografiasyvidas.com/biografia/s/savall.htm [Accessed: December 11, 2024]

[10] Fort MO i. Jordi Savall, Música En'diálogo mayor'. AusArt Journal for Research in Art. 2019;7:243-258. DOI: 10.1387/ausart.20372. Available from: https://www.goldjournal.net/article/S0090-4295(20)30372-1/fulltext

[11] Cooley JW, Tukey JW. An algorithm for the machine calculation of complex Fourier series. Mathematics of Computation. 1965;**19**:297-301

[12] Press W, Teukolsky S, Vetterline WT, Flannery BP. Numerical Recipes: The Art of Scientific Computing, Ch. 12–13. New York, USA: Cambridge University Press; 2007

[13] SciPy tutorial, Fourier Transforms (scipy.fft). 2021. Available from: https://docs.scipy.org/doc/scipy/reference/tutorial/fft.html#ct65 [Accessed: June 14, 2021]

[14] Swarbrick D, Bosnyak D, Livingstone SR, Bansal J, Marsh-Rollo S, Woolhouse MH, et al. How live music moves us: Head movement differences in audiences to live versus recorded music. Frontiers in Psychology. 2019;**9**:60-63. DOI: 10.3389/fpsyg.2018.02682. Available from: https://www.goldjournal.net/article/S0090-4295(20)30372-1/fulltext

[15] Britannica. The Editors of Encyclopedia. "Improvisation". Chicago,

IL, USA: Encyclopedia Britannica; 2022. Available from: https://www.britannica.com/art/improvisation-music [Accessed: June 24, 2022]

[16] Ross SM. Introduction to Probability and Statistics for Engineers and Scientists. 6th ed. Los Angeles, CA, USA: Elsevier Inc.; 2020. DOI: 10.1016/C2018-0-02166-0

[17] Ross S. A First Course in Probability. 9th ed. Harlow, UK: Pearson; 2012. DOI: 10.1016/C2018-0-02166-0

[18] Leemis LM, McQueston JT. Univariate Distribution Relationships. Vol. 62. American Statistical Association; 2008. pp. 45-53. DOI: 10.1198/000313008X270448

[19] Cover TM, Thomas JA. Elements of Information Theory. 2nd ed. Hoboken, New Jersey. USA: Wiley; 2006

[20] Smaldino PE. Measures of individual uncertainty for ecological models: Variance and entropy. Ecological Modelling. 2013;**254**:50-53. DOI: 10.1016/j.ecolmodel.2013.01.015

[21] Freund RJ, Wilson WJ, Mohr DL. Chapter 4 - inferences on a single population. In: Mohr DL, Wilson WJ, Freund RJ, editors. Statistical Methods. Cambridge: Academic Press; 2010. pp. 169-199. DOI: 10.1016/B978-0-12-823043-5.00004-7

[22] Massey FJ. The Kolmogorov-Smirnov test for goodness of fit. Journal of the American Statistical Association. 1951;**46**(253):68-78

[23] Lehoczky JPD. Statistical: Special and continuous. In: Wright JD, editor. International Encyclopedia of the Social & Behavioral Sciences. 2nd ed. Amsterdam, Netherlands: Elsevier; 2015. pp. 575-579. DOI: 10.1016/B978-0-08-097086-8.42115-X

[24] The College of William & Mary. Univariate Distribution Relationship Chart. Williamsburg, Virginia, USA: The College of William & Mary; 2022. Available from: http://www.math.wm.edu/~leemis/chart/UDR/UDR.html [Accessed: July 24, 2022]

[25] Suits BH. Physics of Music-Notes. Houghton, Michigan, USA: Physics Department, Michigan Technological University; 2022. Available from: https://pages.mtu.edu/~suits/notefreqs.html [Accessed: July 24, 2022]

[26] Billard L. Markov models and social analysis. In: Wright JD, editor. International Encyclopedia of the Social & Behavioral Sciences. 2nd ed. Amsterdam, Netherlands: Elsevier; 2015. pp. 576-583. DOI: 10.1016/B978-0-08-097086-8.42144-6

[27] Encyclopedia Britannica. The Editors of Encyclopedia. "Musical Expression". Chicago, IL, USA: Encyclopedia Britannica; 2022. Available from: https://www.britannica.com/art/musical-expression [Accessed: July 25, 2022]

[28] Lugo I, Alatriste-Contreras MG. Nonlinearity and distance of ancient routes in the Aztec empire. PLoS One. 2019;**14**(7):e0218593. DOI: 10.1371/journal.pone.0218593

[29] Lugo I, Martínez-Mekler G. Theoretical study of the effect of ports in the formation of city systems. Journal of Shipping and Trade. 2022;**7**(16):1-16. DOI: 10.1186/s41072-022-00117-6

[30] Lugo I, Alatriste-Contreras MG. Intervention strategies with 2D cellular automata for testing SARS-CoV-2 and reopening the economy. Scientific Reports. 2022;**12**(13481):1-13. DOI: 10.1038/s41598-022-17665-3

Chapter 4

Projecting Event Accrual as a Survival Trial Progresses

Edward Lakatos

Abstract

A critical piece of information for a survival trial is when it is expected to end. This is essential not only for trial planning and execution, but also for funding. The power of a trial is driven by the number of events, rather than patients. Implementing clinical trials presents problems not typically encountered in other experiments, such as with mice. Consider drug compliance. Initially, essentially all patients take assigned drugs. Over time, some patients continue to take all of their drugs all of the time, some take some drug all of the time, some take all of the drug some of the time, etc. Drug discontinuation can affect the required number of events as well as event accrual. Other factors can also affect event accrual, e.g., loss to follow-up or competing risks, staggered entry, delayed treatment effects, etc. The methods presented here use Markov models, originally developed for calculating sample size, but adapted for projecting event accrual. Pre-trial guesstimates of parameters can be updated during the trial to enhance accrual projections.

Keywords: event accrual, Markov model, survival trial, time-dependent, complex clinical trial, projection of events, interim data, Kaplan-Meier, time-varying risk

1. Introduction

Survival Clinical Trials designate a single primary variable for comparing randomly assigned treatment arms. The occurrence of this variable is called an "event" and it can be any dichotomous variable such as death, cancer, heart attack, etc. Survival trials typically end when a target number of events occur, the target number being established prior to the start of the trial. Power is typically the most important consideration for establishing that target.

As with all clinical trials, the size of the treatment effect is a critical factor for establishing power. Many other factors (referred to here as "complexities") can affect the power. Much of clinical trial complexity is due to the fact that patients are usually more difficult to treat and follow, compared with, say, plots of land, or mice. If a patient does not comply with her assigned treatment, the treatment effect is compromised, and so, in turn, is the power. Lakatos [1, 2] developed a Markov Chain model [3] approach for calculating sample size/power in these complex settings, at the design stage. Such calculations necessarily involve guesstimating parameters, such as time-dependent event rates and sizes of treatment effects that will be realized during the trial. If similar trials were recently carried out, for which results are available, those results might be a good source for parameter guesstimates. From a broad perspective, the use of such sources can be considered "data analysis", whether it involves reanalyzing data, or simply gathering and interpreting collections of results from those prior trials.

Since these complex factors can have a substantial impact on power, and in turn, power is central for estimating the target number of events, I will use the methods developed by Lakatos [1, 2] to assess the impact of these complex factors on when the target number of events is likely to be achieved.

This chapter begins (Section 2.1) by showing how published survival curves can be used to extract information for projecting events. Building a Markov model to reproduce the survival curves of Section 2.1 is the topic of Section 2.2, while Section 2.3 deals with including trial complexities into the Markov model. Projecting events prior to the trial is the topic of Section 2.4, and Section 2.5 presents similar projections for when the trial is in progress.

2. Markov model for projecting accrual of events

2.1 Extracting survival information from plots

Figure 1 displays a digital capture program ("GRABIT" [4]) (in MATLAB [5]) in which a plot of published survival curves from a prior (2015) trial [6] ("CHECKMATE 066") has been uploaded. Suppose our new trial will use Nivolumab (Nivo), an

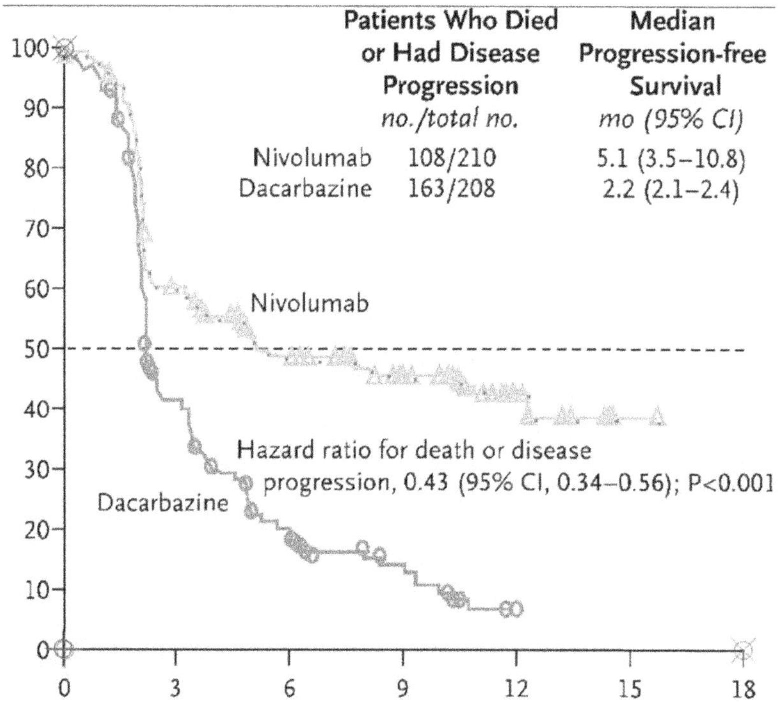

Figure 1.
MATLAB digital capture interface.

immune inhibitor, as the control arm (a new drug will be compared to Nivo). We wish to reproduce these curves in a Markov model (the model will be presented in detail, shortly). The authors [6] estimated the published survival curves in **Figure 1** from raw data using Kaplan-Meier [7] (KM) estimates (details of the KM estimating procedure can be found in most basic textbooks on survival analysis (e.g., [8])). The MATLAB digital capture program requires the user to first calibrate the graph by placing circles at the origin, and ends of the x- and y-axes, as well as entering the coordinates of those three points. Then mouse clicks are used to identify points on the reference curve. To capture the undulations of this curve, many points (mouse clicks) were identified. The KM estimation procedure produces conditional survival probabilities at each distinct event time, and those distinct event times are random. The locations of my mouse clicks (red dots on the Nivo curve (possibly not visible)) were intended to be sufficiently frequent to capture the undulating nature of the curve.

Those captured points were saved to an Excel file and then plotted in **Figure 2** (solid line). The tabular form [8] (i.e., Excel tables) of the KM estimates, rather than the plots, were used in calculations. For compatibility with the Markov model, as well as other factors, it is important to use points defining intervals of uniform width (as opposed to the random widths of the original KM estimates). Thus, the original Nivo KM curve is replaced by the dashed line which has downward steps exactly at each integer month. This approximation to the original KM curves was accomplished using exponential interpolation: first, for each month, identify the closest two time points (on the original KM) surrounding that month, in the tabular form of the KM, and interpolate to find the y-value corresponding to that integer month. Those monthly probabilities are displayed in the first two columns of **Table 1**.

The third column displays conditional survival probabilities $S(t_i|t_{i-1})$, calculated from cumulative survival probabilities $S(t_i)$ in column 2, using

$$S(t_i|t_{i-1}) = \frac{S(t_i)}{S(t_{i-1})} \tag{1}$$

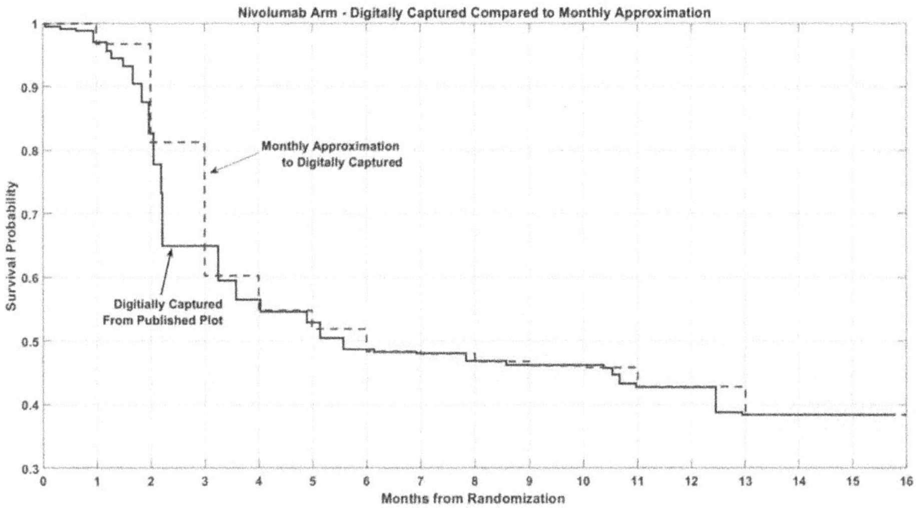

Figure 2.
Digitally captured KM with monthly approximation.

Month	Cumulative Surv Prob $S(t_i)$	Conditional Surv Prob $S(t_i \mid t_{i-1})$	Cond annual fail Prob
0	1	0.967467	0.327592
1	0.967467	0.840246	0.876155
2	0.81291	0.74792	0.969362
3	0.607992	0.901204	0.713003
4	0.547925	0.94768	0.475263
5	0.519258	0.931525	0.57309
6	0.483702	0.991838	0.093661
7	0.479754	0.973838	0.27249
8	0.467202	0.98593	0.156367
9	0.460629	0.993959	0.070133
10	0.457846	0.933535	0.561908
11	0.427415	0.987439	0.140742
12	0.422047	0.91497	0.655744
13	0.38616	0.994133	0.068172
14	0.383894	0.995912	0.047968
15	0.382325	0.995162	0.056533
16	0.380476	1	0

Table 1.
Approximation to KM estimates from digitally captured data, and conditional survival probabilities.

For example, calculating the conditional in (row 2, col. 3) from col. 1:$.81291/.967467 = .840246$. And for the conditional annual failure rate:$1 - (.84021)^{12} = .87622$. The conditional survival rates (col 3) show how the risk changes month by month. And the conditional annual failure rate is more familiar to clinicians: "the failure rate during this month, on a yearly basis is 87.6%". Note that the calculations in (1) are the reverse of the Kaplan-Meier procedure, which calculates cumulative survival probabilities from conditional [8]. This simply reflects the fact that we are extracting information from already established KM curves.

One can also calculate

$$S(t_i) = 1 \times \frac{S(t_1)}{S(t_0)} \times \frac{S(t_2)}{S(t_1)} \times \frac{S(t_3)}{S(t_2)} \times \cdots \times \frac{S(t_i)}{S(t_{i-1})} \tag{2}$$

2.2 Building a Markov model to reproduce the approximation to the KM estimates

The type of Markov Chain model [3] used here is discrete-state, and discrete-time. It is also non-stationary, as the failure rates (etc.) are allowed to vary over time, as, for example, in **Figure 1**. To define this Markov model, is suffices to define an initial distribution and transition matrices. We begin with a simplified Markov model.

Let $\mathcal{D}_{t_i}$ denote the distribution at time t_i, for which there are two states: A for those patients at risk at time t_i for the primary event, and E for those who have already experienced the event. Markov models require these states to be exhaustive and mutually exclusive at any given time. Denote the probability of being in state A at time t_i by $\Pr_{A,t_i}$, and similarly for state E. The mutually exclusive and exhaustive conditions lead to $\Pr_{A,t_i} + \Pr_{E,t_i} = 1$. The notation for this distribution of a two-state model at t_i is

$$\mathcal{D}_{t_i} \equiv \begin{bmatrix} E & A \\ \Pr_{E,t_i} & \Pr_{A,t_i} \end{bmatrix} \tag{3}$$

Here, the letters A and E at the top are not part of the matrix—rather they identify the States. For this Markov model, the initial distribution is

$$\mathcal{D}_{t_0} \equiv \begin{bmatrix} E & A \\ 0 & 1 \end{bmatrix} \tag{4}$$

This indicates that at the time of randomization ($t_0 = 0$), all patients are still at risk with $\Pr_{A,t_0} = 1$, where the "0" in (4) indicates that, at the time of randomization, there are no events, and the "1" indicates that all patients are still at risk at that same time.

Transition matrices for this model are

$$\mathcal{T}_{t_{i-1},t_i} \equiv \begin{bmatrix} & E & A \\ E & 1 & \Pr_{A,E,t_{i-1}} \\ A & 0 & \Pr_{A,A,t_{i-1}} \end{bmatrix} \tag{5}$$

The "1" in the upper left position indicates that events remain events with probability 1, in this model, while the "0" in the lower left position indicates that patients with events never return to the at-risk state (a different Markov model can be defined for Recurrent Events analysis [9]). The probabilities $\Pr_{A,E,t_{i-1}}$ and $\Pr_{A,A,t_{i-1}}$ are called "transition" probabilities. $\Pr_{A,E,t_{i-1}}$ is the probability of "transitioning" from state A to state E at time t_{i-1}, and $\Pr_{A,A,t_{i-1}}$ is the probability of remaining in state A. Because in this model only transitions from state A to state E can take place, $\Pr_{A,E,t_{i-1}} + \Pr_{A,A,t_{i-1}} = 1$.

Given the distribution $\mathcal{D}_{t_{i-1}}$ at time t_{i-1}, the distribution $\mathcal{D}_{t_i}$ is given by

$$\mathcal{D}_{t_i} = T_{t_{i-1},t_i} * \mathcal{D}'_{t_{i-1}} = \begin{bmatrix} & E & A \\ E & 1 & \Pr_{A,E,t_{i-1}} \\ A & 0 & \Pr_{A,A,t_{i-1}} \end{bmatrix} * \begin{bmatrix} E & A \\ \Pr_{E,t_{i-1}} & \Pr_{A,t_{i-1}} \end{bmatrix}' \tag{6}$$

The sequence of distributions $\{\mathcal{D}_{t_i}\}_{t_i=0}^{K}$ can be obtained by successive multiplications of (6), where K is the total number of divisions of the length of the trial. Eqs. (4)–(6) define this general 2-state Markov model.

This Markov model can be used to reproduce the KM curve in **Table 1** as follows. First, at each time t_i, set $\Pr_{A,E,t_{i-1}} = S(t_i|t_{i-1})$, and $\Pr_{A,A,t_{i-1}} = 1 - S(t_i|t_{i-1})$; recall that $S(t_i|t_{i-1})$ was calculated from the digitally captured data in **Table 1**, using (1). Repeated multiplications of (6) lead to a sequence of distributions $\mathcal{D}_{t_i}$, and if these distribution are stacked vertically in a matrix or table, with the row containing the result of the ith multiplication entered in the ith row of the table, then that will "match" the corresponding rows of **Table 1**. We have now reproduced the published Nivo survival curve in a Markov model.

Note that in this Markov model, the event probability is accumulating over time, which is the cumulative failure probability. To obtain the cumulative survival probability, which is presented in **Table 1**, calculate 1 minus the cumulative failure probability. In Section 2.3, in which complexities are added to the Markov model, the probability of being "At Risk" is not necessarily the same as the probability of surviving.

2.3 Expanding the Markov model to address the complexities of clinical survival trials

The 2-state Markov model constructed in Section 2.2 is part of the 4-state Markov model developed for complex clinical trials developed by Lakatos [1, 2]. A more detailed description of those 4 states can be found in [1, 2, 10]. The development in Section 2.2, which emphasized the ability of that Markov model to reproduce KM curves, has not, as of the current time, been presented elsewhere.

This section focuses on the complexities addressed in the 1986 Markov model [1], namely: time-dependent rates of competing risks, loss-to-follow-up, non-compliance, drop-in, and staggered entry. I begin with staggered entry (i.e., patients are enrolled into the trial over a period of time). Staggered entry, (presented originally in [1], but more comprehensively in [11]) is part and parcel of every survival trial, and has direct implications on the rate of event accrual.

To understand the handling of staggered entry requires an understanding of censoring [8]. Censoring refers to a missing data phenomenon of survival trials. For sample size and power calculation, there is a primary variable and corresponding event, such as the incidence of death, or heart attack, or cancer, for which we are studying the survival experience. The underlying survival curve is typically defined for the time period $[0, \infty)$. Most trials last only a few years. At trial's end, there are almost always patients who have not yet had the primary event. For such patients, the time of event will not be available for analysis; their event time is said to be "censored". This type of censoring is referred to as "administrative censoring" because it arises from the administration of the trial (here, the performance of an analysis). Other types of censoring include (1) competing risks: if, for example, the primary event is heart attack, but the patient first dies of cancer; (2) loss-to-follow-up, in which a patient's primary event status, for one of a variety of reasons, cannot be assessed—for instance, the patient relocated to an unknown foreign country. Note that the data from a censored patient is far from useless. In fact, the time of censoring indicates that the censored patient still did not experience the event at that censoring time, so fared better than those with earlier event times.

There are several reasons why administrative censoring differs from other natural censoring phenomenon. The most important modeling reason is that administrative censoring can be calculated precisely from the recruitment pattern. Also, if patients are followed beyond the current analysis, then all times of administrative censoring increase for later analyses. In contrast, when cancer death censors the event time of the primary variable, that censoring time never changes.

Study time (ST) versus time from randomization (RT). When patients are enrolled, times are recorded in calendar time, which are first transformed into time from study start (ST). In many diseases, the risk of failure changes over time. See, for example, **Figure 1**: shortly after randomization (time 0 on the x-axis) both curves plunge steeply, indicating very high risk. In less than 3 months, the Nivo arm risk tapers off dramatically. These Kaplan-Meier curves are all based on time from randomization, transformed from the recorded time from study start. Without the transformation from ST to RT, a Nivo patient's risk at RT time 3 months could be compared to a Dacarbazine patient's at 2 months. The difference due to treatment would be dwarfed by the difference due to different locations in the risk profiles of the two treatments.

When the data are transformed from ST to RT, all patients can be analyzed beginning at the time of randomization, with all sharing the same location in the risk profile of their particular treatment arm.

This transformation assures that the fundamental Markov property is satisfied. In particular, at a given time $t_\nu > 0$ from randomization, all patients still at risk will have advanced the same t_ν units in that common risk profile of their treatment arm (see, again, **Figure 1**), so that the probability of failure will be the same for all these patients. The fundamental Markov property requires that all patients in a particular state, at a given time, have the same risk of failure regardless of how they arrived at that state. This would not be true if the ST time frame had been used, with patients in the same state at the same time having been followed for different lengths of time since their randomization.

Eligibility criteria are designed to recruit patients who are at a similar point in their disease advancement. In order to assure that analyses apply to patients who are at a similar point in their risk profile, all recorded times (study times) are transformed into time from randomization. And randomization is used to assure that the treatment arms are similar.

For this reason, "1" is the probability of being at risk in the initial distribution. It corresponds to the fact that in **Figure 1**, all patients are at risk at time 0 (100% survival).

2.3.1 Implications for administrative censoring

Suppose a trial terminates at a calendar time t_{final}. The ST to RT transformation means that a patient j randomized at study time r_j will have been followed for $t_{\text{final}} - r_j$ time units when administratively censored. The example given in **Table 2** shows the relationship between the randomization pattern and administrative censoring.

In this example, recruitment takes place over 6 months, in the pattern shown in the rows labeled "recruitment period" and "number randomized". The data are analyzed at 180 days, and the total recruited is 590. If n_j is the number recruited in the jth recruitment period, the corresponding probability of being recruited during this period is

$$p_j = \frac{n_j}{\sum n_i} \tag{7}$$

This probability is shown in the "randomization probability" row.

The lower panel, for administrative censoring, is derived from the upper panel. There is a natural correspondence between the recruitment period, j, and the administrative censoring period, m. If K is the number of subintervals of the full trial period

	Recruitment					
Calendar day	**Apr 1**	**+30**	**+60**	**+90**	**+120**	**+150**
Recruitment period j	1	2	3	4	5	6
Number randomized	55	90	85	140	120	100
Randomization probability	0.093	0.152	0.144	0.237	0.203	0.169
	Administrative censoring					
Days: rand to analysis	30^-	60^-	90^-	120^-	150^-	180^-
Admin cens period m	1	2	3	4	5	6
Still at risk at analysis	590	490	370	230	145	55
Number censored	100	120	140	85	90	55
Censoring probability	0.169	0.244	0.378	0.369	0.620	1

Table 2.
Relationship between recruitment pattern and implied administrative censoring.

$[t_o, T]$, then the correspondence is $j \leftrightarrow K - j + 1 = m$. This stems from the transformation from ST to RT discussed above. For example, if a patient is randomized at day 150 ($j = 6$), then that patient will have been exposed to randomized treatment for 30 days at the time of the 180-day analysis; consequently, that patient will be administratively censored at day 30 ($m = 1$). The probability of being administratively censored during the mth period is

$$a_m = \frac{p_{K-m+1}}{\sum\limits_{h=1}^{K-m+1} p_h} \tag{8}$$

2.3.2 Adding a loss state L to the 2-state Markov model to accommodate censoring

The Markov structure differs from the set-up for calculating the KM estimates because of the Markov requirement that the entire distribution (exhaustive) be accounted for at each successive Markov multiplication (6). In the KM setting, we keep track of the events and the patients at-risk; but patients who are censored are not involved in further calculations, so they are dropped from the KM procedure at the time of such censoring. Following Lakatos [1, 2], all patients who are censored transition into the loss state L. This loss state receives patients who are lost for any reason, administratively censored, or otherwise. Similar to the KM procedure, this Markov loss state will not be used for further calculations.

To incorporate staggered entry, a loss state L is added to the 2-State model, resulting in a 3-State model:

$$\mathcal{D}_{t_i} \equiv \begin{bmatrix} E & A \\ \Pr_{E,t_i} & \Pr_{A,t_i} \end{bmatrix} \text{is replaced by} \begin{bmatrix} L & E & A \\ \Pr_{L,t_i} & \Pr_{E,t_i} & \Pr_{A,t_i} \end{bmatrix} \tag{9}$$

$$\mathcal{D}_{t_0} \equiv \begin{bmatrix} E & A \\ 0 & 1 \end{bmatrix} \text{is replaced by} \begin{bmatrix} L & E & A \\ 0 & 0 & 1 \end{bmatrix} \tag{10}$$

and

$$\mathcal{T}_{t_{i-1},t_i} \equiv \begin{bmatrix} & E & A \\ E & 1 & \Pr_{A,E,t_{i-1}} \\ A & 0 & \Pr_{A,A,t_{i-1}} \end{bmatrix} \text{is replaced by} \begin{bmatrix} & L & E & A \\ L & 1 & 0 & \Pr_{A,L,t_{i-1}} \\ E & 0 & 1 & \Pr_{A,E,t_{i-1}} \\ A & 0 & 0 & \Pr_{A,A,t_{i-1}} \end{bmatrix} \tag{11}$$

Administrative censoring can now be added. The administrative censoring matrix is

$$\mathcal{A}_{t_{i-1}} = \begin{bmatrix} & L & \mathbf{E} & \mathbf{A} \\ \mathbf{L} & 1 & 0 & a_m \\ \mathbf{E} & 0 & 1 & 0 \\ \mathbf{A} & 0 & 0 & 1 - a_m \end{bmatrix} \tag{12}$$

Recall that a_m was derived in (8). Administrative censoring is incorporated into to the Markov model as follows [10]:

$$\mathcal{D}_{t_i} = (\mathcal{A}_{t_{i-1}} \mathcal{T}_{t_{i-1}}) \mathcal{D}_{t_{i-1}} \tag{13}$$

Starting with the initial distribution (10), the sequence of distributions

$$\{\mathcal{D}_{t_i}\}_{i=0}^{K} \tag{14}$$

obtained through successive matrix multiplications (13) is used to calculate sample size for the logrank statistic [2].

Note that the "intermediate" distributions in (14) do not have a real world analogue. For example, the initial distribution (10) has all patients initially being at risk; this was discussed earlier and matches the KM curves in **Figure 1**. But this cannot be, because no patients have been enrolled yet. It is only at the last distribution $\mathcal{D}_{t_K}$ that this matches the real world. This is because, as with the KM estimation procedure, time from randomization is an artificial construct; but that construct (not ST time) provides the critical information for understanding the underlying survival phenomenon.

2.3.3 Non-compliance

The impact of non-compliance on power was an issue as early as 1967 [11, 12]. Non-compliance is a complex topic, with many dimensions. Some patients do not comply with their assigned treatment regimens, even when expected to be life-saving. Non-compliance may be due to intolerable side effects, or simply forgetfulness, or something else. While clinical trials attempt to assess whether a treatment works at all, more subtle questions are whether the treatment works if a patient takes, say, 80% of the medication all of the time, or 100% of the medication 80% of the time. For clinical trials, often performed seeking regulatory approval, complex manufacturing methods have not yet been developed for mass distribution. In turn, the variety of pill dose sizes is usually very limited in trials for regulatory approval. The tested dose may be the same for a small woman or large man, as well as anywhere in between.

Dating back to the 1960s, non-compliance assumptions for clinical trials have typically been very simple: at a given time, a patient is deemed either compliant or non-compliant [12, 13].

Assessing non-compliance status is also fraught. Patients are often asked to bring their pill containers with them to clinical visits, so that clinical staff can count the remaining pills. In response, trash receptacles outside clinics are often replete with a plethora of discarded pills. The wife of a deceased trial patient was asked whether her husband took his medicine. "Oh Yes!" she replied, "he was religious about it – every day he took one pill and flushed it down the toilet."

Despite all of these complications, non-compliance is important for calculating power, which affects the rate of accrual of events. A patient cannot respond to a treatment he assiduously avoids. Until the Markov approach, statistical methods for accounting for non-compliance were less than satisfactory.

As far back as the 1960s, statisticians modeled non-compliance assuming that patients could be classified as: (1) those still at risk at time t_i and complying with their assigned treatment regimen, or (2) still at risk, but no longer complying. Compliers are assumed to have one failure rate, while non-compliers, another. For the Markov model, the at-risk state A needs to be divided into two states.

Even with this simplified setting, a complex mixture model is implied. **Figure 3** shows potential transitions of at-risk patients.

These two at-risk groups have different failure rates, so both are being depleted, but at different rates. And as **Figure 3** indicates, a lot more is going on, all affecting the number of patients remaining in the two at-risk States. The "overall" at-risk failure rate is a combination of these two rates. But because of the different rates of depletion, that overall failure rate is constantly changing. And, as implied in **Figure 3**, it is changing in a very complex way. In turn, that time-dependent failure rate is difficult to characterize. Only the Markov approach, which treats this as a mixture, is been able to avoid this pitfall. Because the Markov model assigns separate states to each of the components (A_C and A_N) of the mixture, the failure rate for each of these two components can be specified independently of one another. We are not forced to come up with an "overall" failure rate. And with the simple model of a single rate for compliant, and another for non-compliant, specification is straight forward. Each of the

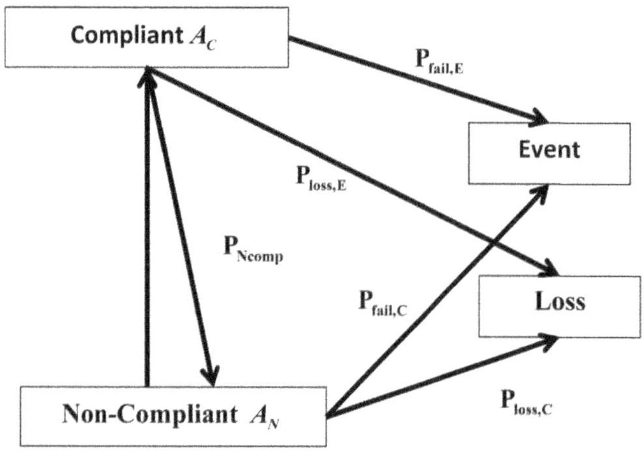

Figure 3.
Compliance flowchart. At-risk compliers, state A_C, and non-compliers state A_N.

arrows in **Figure 3** corresponds to an element in the transition matrix; again, simple to specify.

The 4-state Markov model which includes non-compliance is as follows:

The initial distribution

$$D_{t_0=0} = \begin{bmatrix} L & E & A_C & A_N \\ 0 & 0 & 1 & 0 \end{bmatrix} \tag{15}$$

The transition matrices

$$T_j \equiv T\left(t_j | t_{j-1}\right) = \begin{bmatrix} & L & E & A_C & A_N \\ L & 1 & 0 & p_{\text{loss},j-1} & p_{\text{loss},j-1} \\ E & 0 & 1 & p_{\text{event}_{C,j-1}} & p_{\text{event}_{N,j-1}} \\ A_E & 0 & 0 & 1-\Sigma & p_{\text{dri,j-1}} \\ A_C & 0 & 0 & p_{\text{ncomp,j-1}} & 1-\Sigma \end{bmatrix} \tag{16}$$

Here, $(1 - \Sigma)$ denotes 1 minus the sum of all other entries in the same column. The administrative censoring matrices

$$\mathcal{A}_{t_{i-1}} = \begin{bmatrix} & \boldsymbol{L} & \boldsymbol{E} & \boldsymbol{A_C} & \boldsymbol{A_N} \\ \hline \boldsymbol{L} & 1 & 0 & a_m & a_m \\ \boldsymbol{E} & 0 & 1 & 0 & 0 \\ \boldsymbol{A_C} & 0 & 0 & 1-a_m & 0 \\ \boldsymbol{A_N} & 0 & 0 & 0 & 1-a_m \end{bmatrix} \tag{17}$$

and the matrix multiplications lead to

$$\mathcal{D}_{t_i} = (\mathcal{A}_{t_{i-1}} \mathcal{T}_{t_{i-1}}) \mathcal{D}_{t_{i-1}} \tag{18}$$

It is of interest to note that (18) is identical to (13). It is possible to change the structure of the Markov model (number of States, transition probabilities, etc.) without changing the basic calculating program (18). For example, if the survival distribution by itself is better modeled as a mixture, a State can be added, and the matrices (15)–(17) modified; these matrices reflect the assumptions and structure of the model. The actual mathematical computations are performed using (18).

2.4 Projecting the accumulation of events before the trial

The last distribution $\mathcal{D}_{t_K}$ in the sequence (18) corresponds to the last month of the trial. The end-trial probability of failure for the primary variable is given by p_{event_E,t_K} (see (9)), and for a given sample size N_E, the projected number of events in the experimental group at time t_K is $d_{E,t_K} = N_E p_{\text{event}_E,t_K}$. By varying the trial length t_K, the projected accumulation of events in the experimental group is

$$\{d_{E,t_K}\}_{t_K=1}^{u} \tag{19}$$

The control group is similar.

It is essential to note that the accumulation of events $\{d_{E,t_K}\}_{t_K=1}^{u}$ is not derived from a single evaluation of the Markov model. Rather, each d_{E,t_K} requires Markov matrix multiplications (13) from t_0 through t_K. It is K and t_K that vary. This is necessary to preserve the defining Markov property, that for a given state at a given time, it does that matter how the patient arrived in that state at that time—the transition probability from that state is the same for all patients in that state. As discussed earlier, in order for all patients to begin at the same location in their hazard (or risk) function, the transformation from study time to time from randomization is necessary. In turn, the model assumes that RT time starts at $t_0 = 0$; this is reflected in the initial distribution with the probability of being at risk equal to 1. In the presence of staggered entry, that staggered entry is replaced by modeling the implied administrative censoring.

2.5 Projecting the accumulation of events during the trial

The trajectory of accumulating events in real time is very valuable when the trial is in progress. In early trials (1960 through early 1980s), the length of the trial was posited in months or years from study start (e.g., [14]), and the sample size was posited in terms of the number of patients. Currently, the number of events, rather than the number of patients, determines the power, as well as when the trial ends. While this is true for treatment effects that are constant over time, Lakatos [13] showed that the power of an events-driven trial (i.e., a trial that ends when a pre-specified number of events accrues) can vary dramatically if the treatment effect is not constant.

Even though most survival trials currently are event-driven, trial sponsors need to know the calendar time that a trial is expected to end. This is also critical for investors, and clinics that need to prepare for the next trial. The prediction of that calendar time is the topic of this section.

In Section 2.4, it was shown that, prior to study start, the accumulation of events during the trial could be predicted using the Markov model presented above, together with the formula $\{d_{E,t_K}\}_{K=1}^{u}$ (19) and similarly for $\{d_{C,t_K}\}_{K=1}^{u}$, for the experimental and control arms, respectively. These formulas, together with the per arm sample size, give the number of events in terms of the calendar times t_K, as K increases to u (t_K is the time of a "final" analysis for a given Markov analysis); in (19), t_K can be modified and the Markov model rerun, producing the number of events for that specific t_K.

In Section 2.4, the Markov model is based on pre-trial assumptions, guesstimates which may be off the mark. Suppose, for example, our new trial will use Nivo as the control arm, in a trial similar to the one in which Nivo was shown to be superior to Dacarbazine (the trial known as "Checkmate 066" [6]) (**Figure 1**). The Nivo survival curve in the new trial could still differ from that observed in Checkmate 066. One possibility is that due to variability, the Nivo curve of **Figure 1** may not be a good representation of the true underlying Nivo survival curve. Further, the new trial population, even assuming the same eligibility criteria as Checkmate 066, may be different from those enrolled in Checkmate 066. Indeed, it has been many years since the original was carried out—many of the principle investigators may be new, potentially with new clinical sites (these factors are known to have a substantial impact on study results). Further, the pool of prospective patients might be impacted by the availability of Nivo and other new treatments. There are countless more reasons.

Rather than relying exclusively on the Nivo curves from Checkmate 066, one of many approaches is to explore different assumptions, including other parameters.

In addition, the KM survival curves from the current trial are likely more relevant for predicting the accrual of events in the current trial, than the true underlying survival curves.

The approach presented in this section is to use the survival data from the new trial, possibly in combination with the survival data from **Figure 1**. Obviously, the new trial data will not be fully collected until the trial is complete, too late to be of much use. The approach taken here is to use interim data to estimate the KM survival curves available at the time of the interim. The KM curves for the current trial would then replace the portion of the Nivo survival curve from Checkmate 066 trial for the period for which there is confidence in the new KM estimates.

All of the mathematics for this has already been presented in this chapter. In the remainder of this section, the focus will be on how to replace portions of the pre-trial assumed survival curves with the interim estimated KMs from the current trial. Special programs including interfaces (GUIs—Graphical User Interfaces) were developed (using the MATLAB program "GUIDE" for this purpose. Only one of the interfaces will be discussed. For visual clarity, only sections of the entire interface are reproduced here. Using these sections, I will describe the process the user goes through to obtain event accrual projections.

Begin with a KM curve such as the Nivo curve captured from the Checkmate 066 plot in **Figure 1** (**Figure 4** is from a different project). This curve served as the

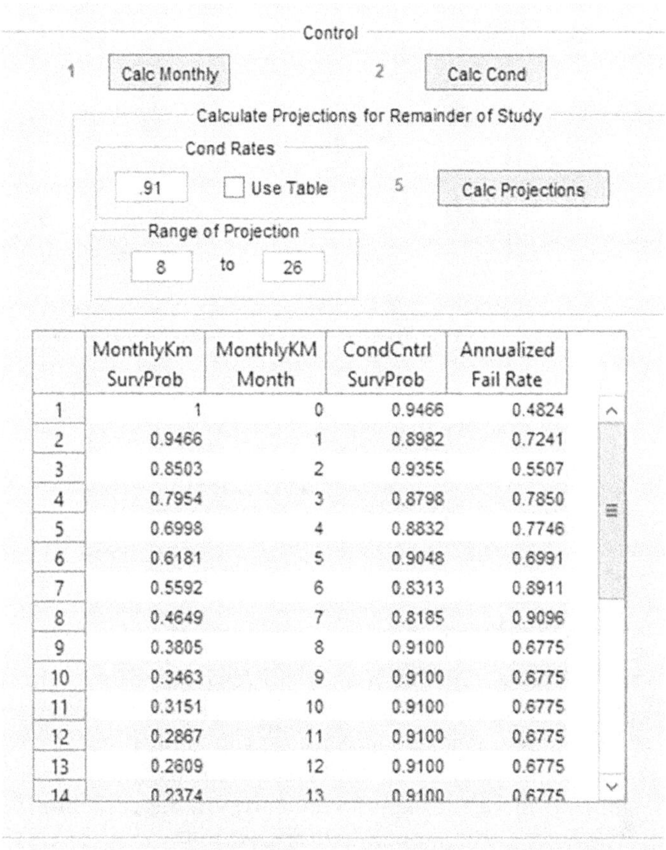

	MonthlyKm SurvProb	MonthlyKM Month	CondCntrl SurvProb	Annualized Fail Rate
1	1	0	0.9466	0.4824
2	0.9466	1	0.8982	0.7241
3	0.8503	2	0.9355	0.5507
4	0.7954	3	0.8798	0.7850
5	0.6998	4	0.8832	0.7746
6	0.6181	5	0.9048	0.6991
7	0.5592	6	0.8313	0.8911
8	0.4649	7	0.8185	0.9096
9	0.3805	8	0.9100	0.6775
10	0.3463	9	0.9100	0.6775
11	0.3151	10	0.9100	0.6775
12	0.2867	11	0.9100	0.6775
13	0.2609	12	0.9100	0.6775
14	0.2374	13	0.9100	0.6775

Figure 4.
Screenshot of portion of interface for setting accrual parameters.

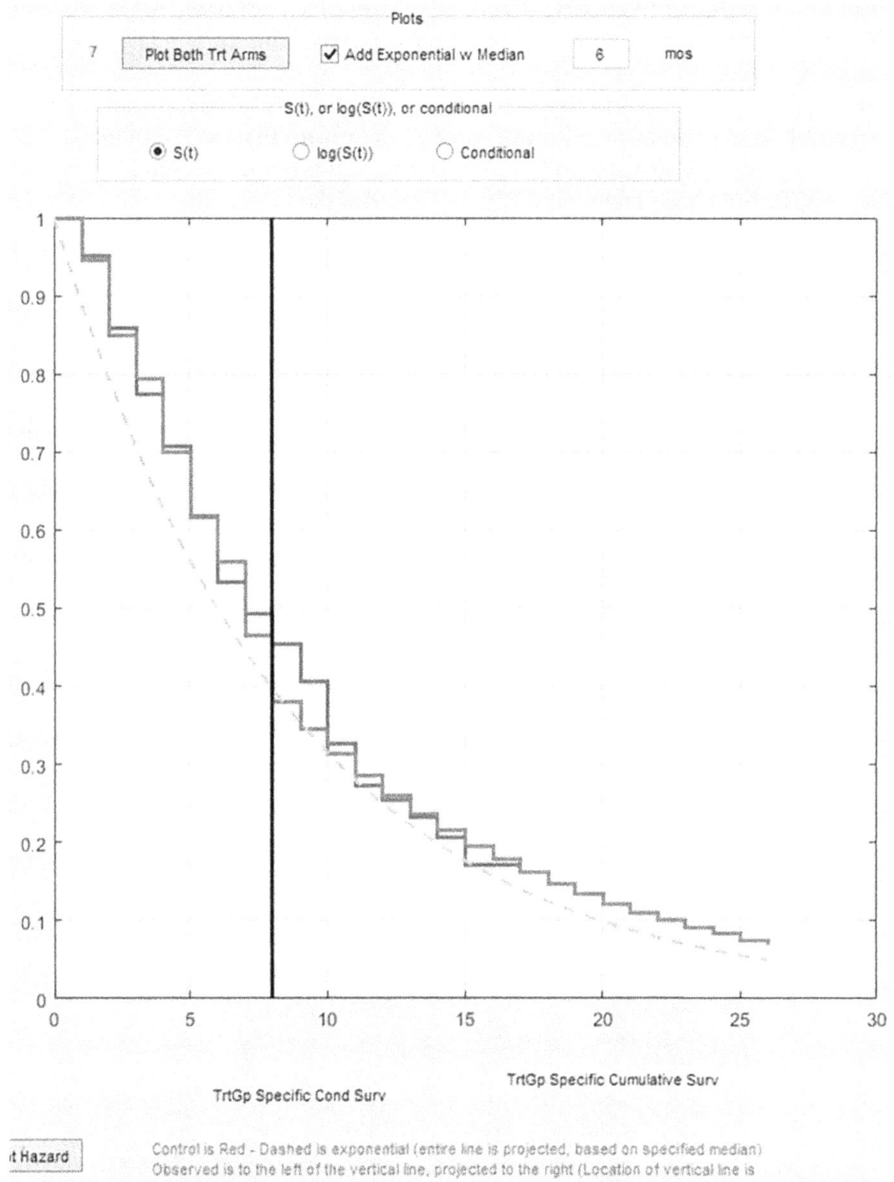

Figure 5.
Screenshot of portion of interface plotting results.

pre-trial survival guesstimate. At this point, the KMs from the new trial have been loaded into the program; after loading, the Table at the bottom of **Figure 4** *is initially blank.* There is a red number next to each pushbutton in the interface, which indicates the sequence of operations performed by the user. The displayed panel is for the control arm. A similar one for the experimental arm, in the same interface is not displayed. The missing consecutive numbers appear in the experimental arm panel.

Pressing the pushbutton "1 Calc Monthly" operates on the loaded data, calculating the monthly approximation to the loaded KM (as was done for **Figure 2**) and enters the results in the first two cols. Pressing the pushbutton "2 Calc Cond" fills cols 3 and 4.

Now we want to subdivide the entire length of the trial into two portions: (1) early portion of the trial, for which we will use the emerging KM data, and (2) the latter portion, for which pre-trial, or other assumptions will be used. This subdivision is user specified under "Range of Projection".

The "Cond Rates" specification, for the latter portion of the trial, allows two choices. Here, .91 was entered. This choice was based either of two choices: "one based on examining the col of the Table "CondCntrl SurvProb", and taking an educated guess, and (2) a KM estimate from a prior study, such as Nivo. In this project, there was no prior study.

Next, the "5 Calc Projections" pushbutton is pushed, and entries in rows 8 through 26 are replaced by the values (.91) the user specified under "Cond Rates". (This could have been KM curves from a previous trial from a table.)

We now switch to another screenshot from the same interface (**Figure 5**). Upon pressing the pushbutton at the top of **Figure 5** "7 Plot Both Trt Arms", the plots dictated in **Figure 4** are displayed in **Figure 5**. Along with the plots of the two treatment arms, is a vertical line at the lower end of the specified range, which demarks the dividing time of the two specifications. In addition, an exponential curve (dashed line) based on the user-specified median (top of **Figure 5**) is produced. The user can change the "type of plot" using the radio-buttons just above the graph. Choices are: "$S(t)$", or "$\ln(S(t))$" or "cond".

The user-designated ".91" was chosen interactively, by specifying a conditional survival probability in the "Cond Rates" panel (**Figure 4**), and observing how that choice changes the right portion of **Figure 4**. All the numbers in Cols 3 and 4 of the table in **Figure 4** are user-editable, so the choice in not limited to the one user-specified conditional rate ".91", or the "Use Table" choice.

Both **Figures 4** and **5** are from a single interface, so that the user can change any specification in **Figure 4**, and immediately see and judge the results in **Figure 5**. The curves on the left and right of the vertical line should be, "visually", relatively consistent. If not, modify the specifications in **Figure 4**.

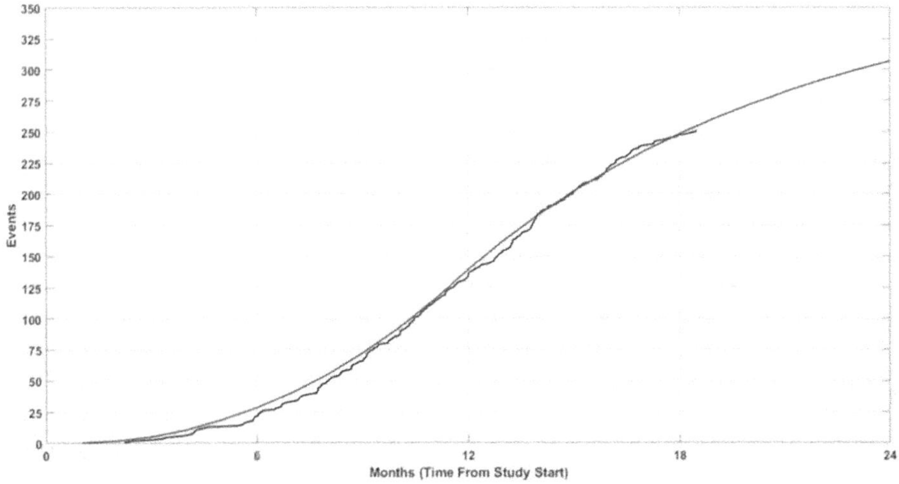

Figure 6.
Projected vs. observed event accrual.

Once the user is satisfied with the choices, those chosen conditional rates as specified in cols 2 (month) and 4(annual fail rate) can be output to an Excel file (pushbutton not shown) and the Excel file loaded into the Markov program designed for producing the accrual of events. **Figure 6** displays a plot of the observed accumulation of events thus far (squiggly line) together with the projected line.

At a particular time (e.g., month 18 from study start in **Figure 6**), the projection is intended to estimate when the desired target number of months will be reached (e.g. 280, or 300, or 350 events). If the projected curve does not reasonably match the observed up to month 18, then the user should modify the projection assumptions in **Figure 4**.

3. Conclusions

The objective of this chapter was to provide methods for projecting the accrual of events for survival trials. The projection methods are based on the Markov model for survival trials [1, 2], and methods described in the earlier part of this chapter. That Markov model was originally developed for calculating sample size and power for survival trials adjusting for complexities routinely encountered in these settings. One of the difficulties is that survival analyses in general are based on time from randomization, while the projection of accrual of events is only useful if those projections are in study time. The factors that typically complicate these calculations involve time-dependent rates of all parameters, for failure, loss-to-follow up, loss to competing risks, non-compliance, non-constant treatment effects, and staggered entry. It was explained why staggered entry differs from other complexities, and a detailed account of how it is modeled was presented. The chapter began by showing how a published KM estimated survival curve can be digitally captured and reproduced as part of a simple Markov model. This simple model was gradually expanded to include adjusting for complexities.

Author details

Edward Lakatos
BiostatHaven, Inc., Croton on Hudson, USA

*Address all correspondence to: ed@biostathaven.com

References

[1] Lakatos E. Sample size determination in clinical trials with time-dependent rates of losses and noncompliance. Controlled Clinical Trials. 1986;**7**(3): 189-199

[2] Lakatos E. Sample sizes based on the log-rank statistic in complex clinical trials. Biometrics. 1988;**44**:229-241

[3] Sekhon R, Bloom R. 10. Markov chains: An introduction. In: Applied Finite Mathematics. Available from: https://math.libretexts.org/Bookshelves/Applied_Mathematics/Applied_Finite_Mathematics_(Sekhon_and_Bloom)/10%3A_Markov_Chains/10.01%3A_Introduction_to_Markov_Chains

[4] Doke J. "GRABIT - extract (pick out) data points off image files"; version 1.0.0.1. In: MATLAB Central. Available from: https://www.mathworks.com/matlabcentral/fileexchange/7173-grabit

[5] MATLAB Version: 9.8.0 (R2017A). Natick, Massachusetts: The MathWorks Inc.; 2017

[6] Robert C, Long GV, Brady B, Dutriaux C, Maio M, Mortier L, et al. Nivolumab in previously untreated melanoma without BRAF mutation. New England Journal of Medicine. 2015; **372**(4):320-330

[7] Kaplan EL, Meier P. Nonparametric estimation from incomplete observations. Journal of the American Statistical Association. 1958;**53**(282): 457-481

[8] Machin D, Cheung YB, Parmar M. Survival Analysis: A Practical Approach. 2nd ed. Chichester: Wiley; 2007. 278 p. DOI: xxx/97804787040-2

[9] Cook R, Lawless J. The Statistical Analysis of Recurrent Events. New York LLC: Springer-Verlag; 2007. 424 p

[10] Lakatos E. Designing complex group sequential survival trials. Statistics in Medicine. 2002;**21**(14):1969-1989

[11] Schork MA, Remington RD. The determination of sample size in treatment-control comparisons for chronic disease studies in which drop-out or non-adherence is a problem. Journal of Chronic Diseases. 1967;**20**: 233-239

[12] Halperin M, Rogot E, Gurian J, Ederer F. Sample sizes for medical trials with special reference to long-term therapy. Journal of Chronic Diseases. 1968;**21**(1):13-24

[13] Lakatos E. The Markov model for survival trials. In: Chen D-G, editor. Biostatistics in Biopharmaceutical Research and Development: Clinical Trial Design. Vol. 1. Cham, Switzerland: Springer; 2025. pp. 19-82. DOI: 10.1007/978-3-031-65947-8

[14] Probstfield JL, Applegate WB, Borhani NO, Curb JD, Cutler JA, Davis BR, et al. The systolic hypertension in the elderly program (SHEP): An intervention trial on isolated systolic hypertension. Clinical and Experimental Hypertension. Part A: Theory and Practice. 1989;**11**(5–6): 973-989

Chapter 5

Future Prediction through Planned Experiments

Tanvir Ahmad and Muhammad Aftab

Abstract

In the recent data-driven world, the ability to predict future results using experimental data is an appreciated work. This chapter explores the concepts of predicting future outcomes from a controlled experimental process, studying both experimental design and analysis techniques for accurate predictions. A well-planned experiment is crucial for attaining reliable data to accurately represent the characteristics of the population under study. We have discussed about classical design structures as well as irregular designs, and the strengths and limitations of each. Furthermore, the well-observed experimental data is analyzed for prediction purposes. Techniques such as; regression analysis, correlation analysis, hypothesis testing and advanced machine learning techniques are used while predicting unknown statistical models. Furthermore, we have explored the implications of model overfitting on predictions and have presented solutions to improve model performance. The role of experimental design for tuning of hyperparameters for one of the machine learning techniques has also been incorporated. This chapter presents a comprehensive examination of how experimental data can be used to make future predictions. Through a combination of theoretical concepts and practical examples, readers will gain a sound understanding of the predictive process for reliable decision-making and policy-making in real-world scenarios.

Keywords: artificial neural network modeling, hyperparameter tuning, optimization, over-fitting, predictive models, response surface designs

1. Introduction

Well-planned experiments are fundamental to collecting systematic, high-quality data. Such methodically gathered information is essential for understanding population characteristics, making predictions, and informing decision-making processes. When researchers analyze experimental data, they can draw meaningful conclusions and make reliable projections about future outcomes. Through careful sampling techniques, researchers can gather data efficiently and cost-effectively while still obtaining meaningful insights about the broader population. However, the key to reliable results lies in ensuring that the selected sample accurately represents the whole population without bias. In our research, we specifically employed response surface design methodology to generate reliable sample data through controlled laboratory experiments.

2. Experimental design to generate sample data

Process optimization involves refining both model parameters and their corresponding outputs. Since outputs are typically unknown, polynomial models are used to approximate the relationship between input variables and responses. Response Surface Designs (RSDs) are essential tools that generate controlled experimental data by optimizing input factors across the experimental space. Through well-designed experiments based on Design of Experiments (DOE) principles and response surface methodology, researchers can systematically manipulate independent variables to observe their effects on dependent variables, using both traditional and innovative RSD structures.

3. Response surface methodology (RSM)

RSM is comprised of collection of methods useful for modeling and prediction analysis about problems in which response of interest is influenced by set of input variables and the objective is to optimize the response. Since the response function is unknown, a well-planned model is searched to approximate the unknown function. While the statistical techniques and variables are well-defined, the specific nature of their relationships in complex systems often requires deep exploration. RSM is employed to model and optimize these relationships among variables under study. The analysts' task is to select an appropriate model and estimate its parameters to best describe the underlying relationships. Later on Analysis of Variance (ANOVA) is employed to find the significant contribution of the study factors.

4. Response surface designs (RSDs)

Experimental plans mainly designed for exploring response surfaces are known as RSDs. These designs assemble the observations of response variable to estimate p number of unknown parameters i.e. $p = (k + d)!/k!d!$ of the response surface model of order d with k input factors. The choice of design depends upon the order and structure of the assumed model being used to approximate the surface. To approximate the unknown response function, first and second order RSDs, corresponding to first order and second order models, are mostly being studied. The major issue in the choice of RSD is to select n trials which, under some definite criteria, are suitable for the estimation of the p unknown parameters of the response surface model. For the proper estimation of the model, the design points (trials) should be at least as many as the number of unknown estimates.

4.1 First order RSDs

Mostly in RSM, the type of the relationship between output and set of the input variables is not known. Therefore, the first and most important step in RSM is to search an appropriate approximation for the true functional relationship between the dependent variable and the set of independent variables. Generally, a polynomial of low order in some space of the input variables is used. The approximating function via the first order model incase if the response is well modeled by a linear function of the independent variables is presented as:

$$Y = \beta_0 + \beta_1 x_1 + \beta_2 x_2 + \beta_3 x_3 + \ \dots \ + \beta_k x_k + \varepsilon \tag{1}$$

where $x_i, \ \dots, x_k$ are the coded levels of the k factors; β's are the undetermined parameters and $b^{'}s$ are their estimates. Full factorial designs (FDs), fractional factorial designs (FFDs) and Placket Burman Designs (PBDs) are generally used as first order RSDs. These designs have two levels of each factor. These designs have the capability to estimate, the main effects and the interaction terms. The detail of some of the first order RSDs is given below.

4.1.1 Factorial designs

Factorial designs (FDs), denoted as 2^k, are the first order RSDs. These allow the experimenters to examine the effect of independent variable(s) on an output variable and to determine the probable interaction(s) of several independent variables. However, with the increase of number of input variables, the design size increases exponentially.

4.1.2 Fractional factorial designs (FFDs)

In FDs, the design size increases rapidly with the inclusion of each input factor. To solve this problem, scientists generally use FFDs. FFDs are the reduced version of the full factorial design, in which only a part of the design runs is employed. This design structure of FFDs uses the resources more efficiently, with a tradeoff in information gained. It allows estimating the main effects of each input factor, but not all the interaction effects. Since few interaction effects are confounded or aliased with other effects, meaning that they cannot be separated or distinguished statistically. The standard notation for 2 level FFDs is 2^{k-p}; where 2 are the levels of each input factor (k) and p is the size of the fraction.

4.2 Second order RSDs

The second order polynomial model has the capability to address most empirically observed relationships between inputs and outputs in various processes and systems, as shown in Eq. (3).

$$Y = \beta_0 + \sum_{i=1}^{k} \beta_i x_i + \sum_{i=1}^{k} \beta_{ii} x_i^2 + \sum_{i=1}^{k-1} \sum_{j>1}^{k} \beta_{ij} x_i x_j + \varepsilon \tag{2}$$

In estimated form:

$$E\ (y|x) = b_0 + \sum_{i=1}^{k} b_i x_i + \sum_{i=1}^{k} b_{ii} x_i^2 + \sum_{i=1}^{k-1} \sum_{j>1}^{k} b_{ij} x_i x_j \tag{3}$$

where $x_i, \dots, x_k$ are the coded levels of the k factors contributing for the response y; β's are the regression coefficients. A rich class of second order RSDs, called subset designs, has been presented by Gilmour [1]. Afterward, Ahmad and Gilmour [2] and Ahmad et al. [3] extended this work and introduced many subset designs by exploring the experimental region in variety of ways. Subset designs also include commonly used designs such as central composite design (CCD) and Box Behnken design (BBD).

4.2.1 Central composite design (CCD)

Box and Wilson [4] introduced CCDs. CCDs are the sequential-natured designs and are the basis of the sequential assembly of the RSDs. These are mainly five level designs, containing full factorial or fractional factorial designs of Resolution-V denoted as n_f for the k-factors, 2 k axial or star points as a second part and finally n_c center runs. Factorial or corner points are used to find the estimates of main effects (linear) and interaction effects (bi-linear). Axial or star points are used to estimate quadratic effects and center runs provide extra *df* to estimate the pure experimental error.

A two-variable pictorial layout of CCD with $n_f = 4$, $n_\alpha = 4$ and $n_C = 4$ is presented in **Figure 1**, where FD points reside on the corners of the cube, center runs at the center and axial design points on the axis. The value of α in the axial points is determined by $(n_f)^{1/4}$ for a rotatable design, $\sqrt{k}$ for spherical region and 1 for cuboidal region of experimentation.

This concept of CCDs can be extended to higher number of factors. For more than five factors, fraction of Resolution-V is employed.

4.2.2 Box–Behnken design (BBD)

Second order polynomial model requires at least three levels of each of the k independent variables under study. Occasionally due to economic or ease of experimentation, we prefer to have three levels of each factor. Box and Behnken [5] have

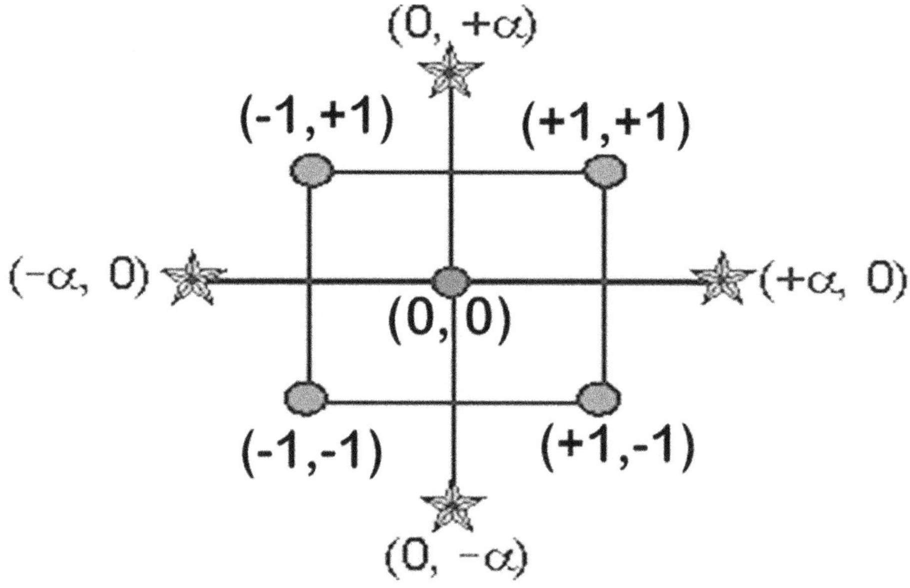

Figure 1.
CCD for k = 2 with center points. Figure copy link: https://www.statease.com/docs/v23.1/designs/ccd/

introduced an important class of three level designs for the estimation of the parameters in the second order model. For scientific studies where RSM is required, if researchers desire three equally spaced levels, the BBD is an efficient alternative to the CCD. The number of design points for $k = 3$, 4 and 5 are quite comparable to CCD. For $k = 3$, the CCD contains $14 + n_C$ (8 factorial, 6 axial and some center runs) design points whereas BBD (**Figure 2**) contains $12 + n_C$ runs. For $k = 4$, the CCD and BBD both contain $24 + n_C$ design points. For $k = 5$, the CCD contains $42 + n_C$ (32 factorial, 10 axial and some center runs) design points whereas BBDs contain $40 + n_C$ runs

BBDs have practical advantages due to the requirement of only three levels for each factor. These designs are not strictly cuboidal or spherical in nature but designs are commonly used for response optimization. Mostly, the designs are rotatable or nearly rotatable, and are suitable to run the experiment in blocks and the blocks are orthogonal to each other. The BBDs are constructed on the notion of balanced incomplete block design Additionally, the center points must be added equal in number among the blocks in order to achieve the orthogonality [6, 7].

5. Experimental region of interest

It is the area where the experimenter is interested to perform the experiment. This area is determined by the ranges of the quantities of the factors, the experimenters agree to run the experiment. Sometimes the researcher is interested to explore the whole experimental region which is known as the complete operability region. Sometimes it is not possible due to the shortage of the resources. Operability region which is completely enclosed inside the limited location of interest is discovered only. The common experimental regions of interest are of two types either cuboidal or spherical. However, the region where the experiment is conducted cannot be limited.

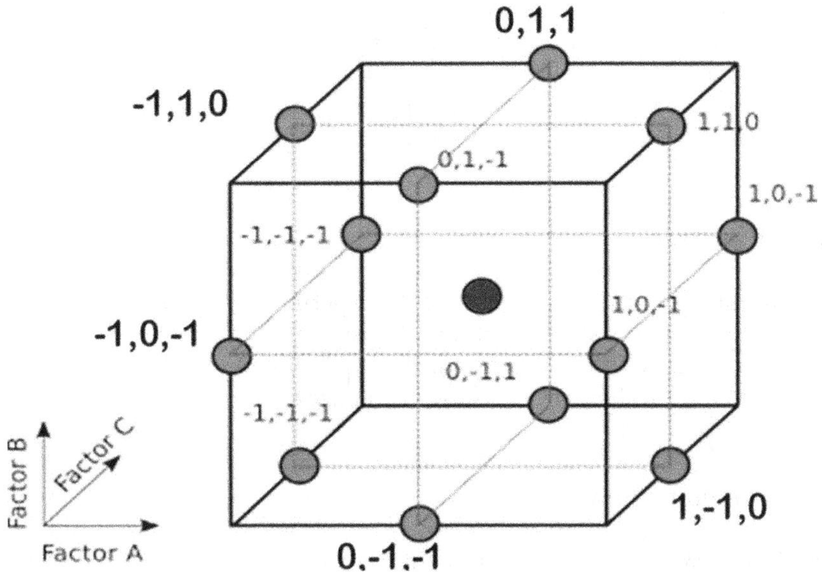

Figure 2.
BBD for k = 3 with center points. Figure copy link: https://www.sciencedirect.com/topics/engineering/box-behnken-design

5.1 Cuboidal region

Cuboidal region of experiment includes the experimental region which contains edge points, points on the face of a cube or hypercube, the vertices points and center points. A face-centered CCD is a familiar second order RSD is which factorial points are placed at corners while center points are located at center of cube/hypercube and axial distance is fixed at $\alpha = 1$ is a cuboidal design. Because of the simpler structure and less number of levels of factors, cuboidal designs are common among the users.

5.2 Spherical region

Spherical region of experiment is based on design points that are assigned on a sphere or hyper-sphere of radius usually equal to $\sqrt{k}$. A CCD is assumed to be spherical when all factorial and axial points are located at the same distance from the center of the design. Spherical CCDs are obtained by setting $\alpha = \sqrt{k}$, where k are the number of factors to be studied.

6. Error in model fitting

After model fitting, the researchers keenly observe the difference between observed and fitted values, known as error. If the observed and fitted values are very close to each other then error value will be small otherwise high error value will be determined. In experimental design data errors are of two types, i.e. pure error and lack-of-fit errors.

6.1 Pure error

It is based upon the replication or repeating of the experiment under identical conditions and it provides information about the internal variation of the data. Generally, performing the experiment at the center is easier because it runs the experiment at the middle levels of each factor, without disturbing the orthogonality of the design. To estimate pure error, replication of the design at the center is the most common practice.

6.2 Lack-of-fit error

It is determined by the unique trials of the experiment. Unique trials are used to estimate the model terms and the extra trials provide degrees of freedom to test the lack-of-fit of the model. Most of the innovative subset designs used in the present study have provided *df* for both lack-of-fit and pure error.

7. Model adequacy measures

Statistically after building and fitting the model, the next step is to check the validity and adequacy of the model. There is a list of methods available which provide information about the reliability and adequacy of the model. Coefficient of determination R^2 and adjusted R^2 were used to judge the adequacy of the performed model.

The higher the value of R^2 i.e. close to 1, the better the fitted model will be. However, R^2 should be used with attention because R^2 may keep increasing with the addition of more terms to the model, which does not necessarily prove the significance of that term [8]. Adjusted R^2 measure is used to hold the effect of variable addition in the increase of R^2.

8. Response plots

Two graphs are typically generated using some statistical software like R, Minitab, Design Expert, JMP etc. by the researchers using RSM known as contour plot and 3D response plot.

8.1 Contour plot

A contour plot is a graphical way of showing a 3-dimensional surface by plotting constant response Z slices, called contours, on a 2-dimensional layout (**Figure 3**). That is, given a value for Z, lines are drawn for connecting the (X, Y) coordinates where the Z value arises. The contour plots reveal the relationship between 2 input variables and a dependent variable. The X and Y values are displayed along the X and Y-axes and contour lines and bands represent the Z value. The graph shows values of the Z variable for combinations of the X and Y variables. Thus contour lines are produced to show the connected points that have a constant response.

8.2 3D response surface plot

Alternative to contour plot, a 3-dimensional response surface plot some time offers a more intuitive visualization, to better understand how the response varies across the entire experimental space (**Figure 4**). Both contour and surface plots help to identify the constant responses in relation to varying input values.

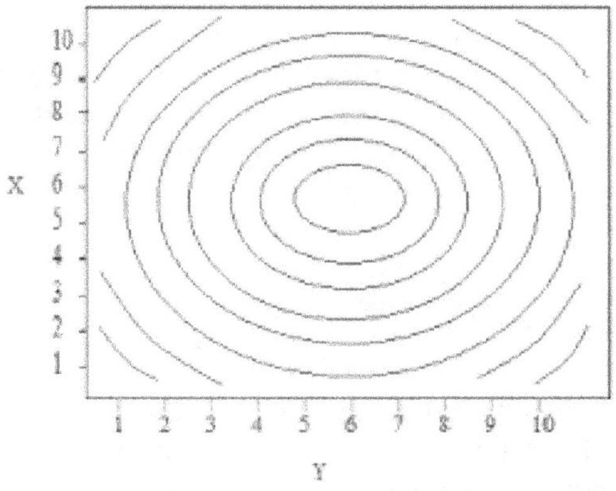

Figure 3.
Display of contour plot. Figure copy link: https://www.itl.nist.gov/div898/handbook/eda/section3/eda33a.htm

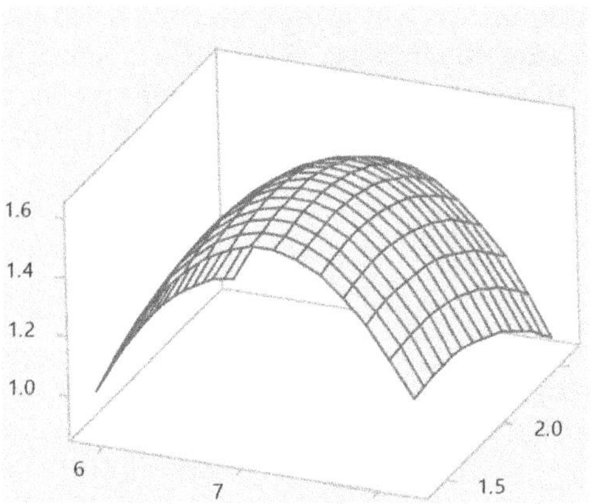

Figure 4.
Display of 3D response surface plot.

9. Artificial neural network algorithm

The term "artificial neural network" (ANN) is derived from biological neural networks (NNs) which perform like the human brain. Like the signals attained by the brain of a human, ANN has nodes that are connected to each other in different layers of the networks [9]. It has a remarkable capability to find sense from complex data and can be applied to find patterns and identify trends that are too difficult to observe either by human beings or computer algorithms [10]. ANN is comprised of three types of layers; input layer(s): receive initial data, hidden layer(s): processes the received data to make useful information and output layer(s): produces final results. A feed-forward network is comprised of units that have one-way connections to other units arranged in layers. Connections only move through the layers. A typical feed-forward network can be represented by the function:

$$y_k = \varnothing_0 \left[\alpha_0 + \sum_{all\ h} w_{h0} \varnothing_h \left(\alpha_h + \sum_{all\ i} w_{ih} x_{ki} \right) \right] \tag{4}$$

where x_{ki} represents the ith value of the kth input vector corresponding to the kth response y_k. Parameter α_h means the connection weights between the constant input and the hidden nodes and α_0 denotes the weight of the direct connection between the constant term and the output. The values w_{ih} and w_{ho} denote the weights for the other connections between inputs and the hidden nodes and the output layers respectively. The functions φ_h and φ_0 denote the activation functions used at the hidden and output layers [11].

9.1 Activation function

The activation of the neurons depends upon the activation functions. The basic objective of the activation function is to introduce non-linearity into the output of a neuron. Activation functions for the hidden layer control the working of the network model from the training dataset. For output layer, it defines the prediction classes of the model. There are some commonly used activation functions. Sigmoid activation function has an S-shape, representing its differentiability and ability to provide a smooth gradient without any abrupt changes in output values (**Figure 5a**). Mathematically it can be expressed as:

$$f(x) = \frac{1}{1 + e^{-x}} \tag{5}$$

The Tanh activation function shares similarities with the sigmoid function, as both have an S-shape. However, the sigmoid function outputs values between 0 and 1 and the Tanh function has a range of -1 to 1 (**Figure 5b**). Mathematically it can be expressed as:

$$f(x) = \frac{(e^x - e^{-x})}{(e^x + e^{-x})} \tag{6}$$

Rectified linear unit (ReLU) activation function provides an impression of a linear function. ReLU has a derivative function and permits for back-propagation which altogether makes it computationally effective (**Figure 5c**). Mathematically the function can be represented as:

$$f(x) = \max(0, x) \tag{7}$$

9.2 Description of hyperparameters of ANN model

Hyperparameters of the ANN algorithm enhance the model performance. These are many in number like; hidden neurons, learning rate, batch size, epochs, momentum rate, etc. Hidden layers are the layers between input and output layers. Many neurons in the hidden layer may increase the complexity of the model and may cause model over-fitting. Learning rate (LR) determines how rapidly a network updates its parameters. The low value of LR slows down the process of learning but converges efficiently. A large value of LR gears up the process of learning but it may not converge. A number of epochs is the number of times the whole training data is shown to the network while training. Batch size is the part of samples given to the network after which parameter values are updated. Commonly batch sizes of 32, 64, 128, 256, ... are being used for very large data sets. For small data sets batch size with small values should be employed.

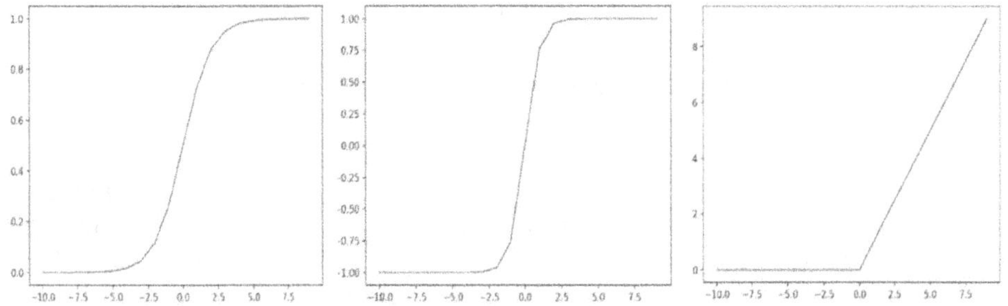

a: Sigmoid function b: Tanh function c: ReLU function

Figure 5.
(a). Sigmoid function. (b) Tanh function. (c) ReLU function.

9.3 Choice and selection of hidden layer(s) and nodes

The most suitable determination of hidden layers and neurons is not easy. The best number of the hidden layer(s) and nodes are those on which the NN produces a low error without being over-fitted. Trial and error technique along with grid search methods is currently being used [12]. Hecht-Nielsen [13] observed that one hidden layer with exactly $2n + 1$ nodes can be used to approximate any function, where n is the input node. Haung [14] said that the size of required hidden nodes can be up to the number of training observations. The standard multilayer feed-forward networks with only one hidden layer have the capability of universal approximation [15]. Berke and Hajela [16] suggested that the number of nodes should be somewhere from the average or sum of the input and output nodes. This way of deciding about hidden nodes will provide ANN model approximation order somewhere between that of a first and second order polynomial. Nodes in hidden layers determine the order of the polynomial, less number of nodes indicates low order polynomial and large number of nodes indicates higher order polynomial approximation [17]. Sheela and Deepa [18] suggested the following formula to determine hidden nodes:

$$N_h = \frac{(4n^2 + 3)}{(n^2 - 8)} \tag{8}$$

where n is the inputs. For small data sets, especially in laboratory based chemical process, more hidden neurons may cause model over fitting and should be avoided.

9.4 LASSO regularization technique to restrict over-fitting

The over fitting problem which arises when ANN technique is applied on small data sets. One popular regularization tool to resolve this issue is using least absolute shrinkage and selection operator (LASSO. Mathematically it is expressed as:

$$L_1 = \min\,(y - \hat{y}(\beta_i))^2 + \lambda \left| \sum_{i=1}^{p} \beta_i \right| \tag{9}$$

where λ is the tuning parameter which controls the bias-variance tradeoff and estimates its best value via cross-validation method [19]. This approach is particularly useful to eliminate the insignificant weights, i.e. these are shrunk to zero.

9.5 Non-linear approximation order and over-fitting issues

In **Table 1**, two of the commonly ignored problems have been highlighted. First, the order of the approximation is generally being overlooked while comparing the efficiency of both RSM and ANN models. These are a few reference articles that are cited and it is clarified that there is no comparison of RSM and ANN models as both have different approximation orders. RSM generally assumes second order polynomial model approximation but the ANN model considers higher order approximation by using more hidden neurons [17]. Second, for more hidden nodes, there are generally fewer observations in the laboratory process to estimate ANN model unknowns. In ML models, this is the case of over-fitting which is again being overlooked. In this

Reference articles	Inputs used	Hidden neurons	Design structure(s)	Data size	Testing data	RSM unknowns	ANN unknowns
[20]	4	20	CCD	30	10	15	80
[21]	4	14	BBD	27	8	15	56
[22]	2	10	CCD	13	5	6	20
[23]	4	10	CCD	30	9	15	40
[24]	3	8	BBD	17	5	10	24

Table 1.
Over-fitting problem cited from the literature.

problem, the model works better for training data but not well for the testing (unseen) data and starts memorizing the results rather than generalizing.

10. RSM and ML techniques: Opinions of the researchers and scientists

After the data collection, ML algorithms are used to optimize the response. Prior to experimentation, the task of proper designing of the experiment is most important. There is an essential role of RSM in the network structure of the deep learning model. Also, RSM plays a significant role to search for the optimal values of the input factors among many. Despite that still some of the researchers have created a sort of conflict by declaring one technique as a winner by using such confiscatory concluding remarks like "ML has outperformed RSM, ML is superior to RSM, ML is the replacement of RSM, etc." which is not the case, as revealed in the literature.

Many big names from Mathematics, Statistics and Data Science have said that Statistics and ML techniques should be used as complementary tools. A mathematician Shafer [25] said Statistics and Artificial Intelligence (AI) have many similarities. Statistics and AI both should learn from each other in spite of their distinctions and differences. Balkin and Lin [26] concluded that NNs cannot replace linear regression as a statistical technique but should instead be considered an additional method for fitting response surfaces. NNs can be used for discrimination and classification but no one is the replacement of the other [27]. ANN has a close resemblance to the famous statistical methods of designed experiments, regression analysis, and RSM but definitely not a replacement for them. This is because NN is a predictive model and cannot provide basic understanding of the underlying process mechanism that produced the data [28]. ANN in Statistics can play a vital role in classification, pattern recognition, prediction and optimization [10]. A well-known data scientist Brownlee [9] perceived that in ML there is a need of Statistics at each step of a project. Statistical methods must be considered as essential part of AI systems, from the preparation of the research questions, designing and planning of the experiment, from the analysis up to the interpretation of the findings [29].

11. Approximation order of RSM and ML techniques

In the recent research era, there has been substantial use of RSM and ML tools (especially ANN) for prediction and optimization purposes. In RSM, second order polynomial approximation and in ML, polynomial approximation without any limit is used to estimate the unknown functions [17]. ANN considers nonlinear approximation up to any order with at least one hidden layer and nonlinear activation functions [15]. ANN is generally compared with second order polynomial approximation by many researchers which has no justification.

In the current study, the nonlinear approximation orders of ANN have been equated to the polynomial approximation order used for RSM to validate the prediction comparison. In many experiments where cost is the major point of consideration, various conventional and unconventional RSDs are used for the process optimization. This may save the resources of the experimenter and may better explore the experimental region.

12. ANN modeling: Role of the hidden layers and neurons

In the ANN technique, hidden layers and their nodes have a key role to approximate the order of the polynomial model. More number of hidden layers and nodes has higher order polynomial model approximation. While comparing the prediction capability of RSM and ANN, researchers have used hidden layers and nodes without knowing its significance for model approximation order. A comprehensive review about the selection of hidden layers and neurons is under study. This review will provide guidelines to the researchers to select the appropriate values of hidden layers and neurons. Suitable selection of the middle layer and their neurons will make the model approximation order same for ANN to the RSM.

13. RSDs to tune hyperparameters

ML techniques have parameters along with hyperparameters in their models. Appropriate selection of hyperparameter values helps to enhance the model performance. Optimal selection of the hyperparameter values enhances the model performance. Better tuning of hyperparameters permits the researchers to enhance model performance to attain optimal results. Commonly, researchers have considered trial and error method, grid search, random search, constructive and pruning methods to choose the values of the hyperparameters. In some studies, Factorial Designs (FDs), CCD and BBD of RSM have also been used for the selection of optimal values of hyperparameters. In the present work, conventional and new proposed RSDs are used to tune the hyperparameters values of ANN model.

14. RSM and ANN modeling for the process optimization

Balkin and Lin [26] discussed that RSM is concerned with estimating a surface to a particularly small set of samples. The basic objective is to optimize the response by determining the appropriate levels of the input variables. Usually, second order polynomial model efficiently approximates the unknown response function. ANNs are recognized to be universal function approximators under specific conditions. This remarkable ability to approximate undetermined functions to any desired degree of accuracy makes ANNs an attractive tool for use in a response surface analysis. The authors have presented ANNs as an additional in RSM and presented their use theoretically and practically.

RSM and ANN are most frequently being used in process optimization nowadays. It is very important to plan the experiment properly so that reliable data can be obtained. We employ RSM for experimental design, followed by a comparative analysis of predictive performance using both RSM and ML tools, with specific focus on ANN. Researchers have used this methodology to improve the optimization performances via selecting optimal values of the input factors for the optimization of the response (see [22, 29–38]).

14.1 Application in textile industry

The textile industry has increasingly adopted eco-friendly natural dyeing processes in recent years. In this work, cotton fabrics have been dyed through hulela zard (*Terminalia Chebula L.*) plant in a laboratory experiment. Innovative subset designs have been used to optimize the important dyeing parameters, namely pH, salt, temperature and time, for optimizing the dyeing performance. The operating ranges of these independent parameters are extract pH (5–9), salt for exhaustion (1–3 g/100 mL), dyeing temperature (50–90°C) and dyeing time (30–70 min). K/S value was observed as a response variable; where K/S is a value used to determine the depth of color in a dyed fabric, and it stands for the absorption coefficient (K) and scattering coefficient (S).

14.2 Analysis of sample data obtained by CCD

Table 2 presents the experimental results. It shows that main effects such as: *p*H, salt and time play a significant roles in the response. At the same time, *p*H*Temperature and *p*H*time are the two-way interaction terms of the dyeing variables that were found significant. The *F-value* of 23.93 and low probability value [*p-value* = 0.000] implies that the model was highly significant. The R^2 = 0.9571 which is remarkably high, shows that only 4.29% of the whole variance was not explained by the given model. A low value of S.D. = 0.1274 determines the excellent experimental performance.

14.3 RSDs contribution for tuning hyperparameters

Complex predictive models are prevalent in the published research. Commonly high values of the hyperparameters such as batch size and hidden neurons are set by the researchers which make the model unnecessarily complex and time consuming. Hyperparameters shown in **Table 3** with batch size (2–10), hidden neurons (1–5) and epochs size (5–25) were investigated by using new RSDs. The natural levels of the hyperparameters were converted as high, low and middle with +1, −1, and 0 levels as codes respectively. RSDs have taken less number of runs to search for the optimum values of these hyperparameters in comparison to customarily used methods such as random search and grid search. By considering RSDs, optimal values for these hyperparameters were found as; batch size = 2, number of epochs = 15 and hidden nodes = 3 (**Figure 6**). It is clear that better model adequacy (above 96%) has been obtained by considering the values of these hyperparameters.

14.4 Modeling with ANN methodology

In the present study, the ANN model structure of (4-3-1), i.e. four input dyeing parameters, three hidden neurons, and the K/S value in the output layer has been employed. The ReLU as a transfer function and SGD as an optimizer were used.

Source of variation	*df*	Adj. SS	Adj. MS	F-value	*p*-value
Model	14	5.43665	038833	23.93	0.000
Linear	4	0.28330	0.0783	4.36	0.015
*p*H	1	0.09134	0.09134	5.63	0.031*
Salt	1	0.11214	0.11214	6.91	0.019*
Temperature	1	0.00007	0.00007	0.00	0.947
Time	1	0.07975	0.07975	4.91	0.042*
Square	4	4.88373	1.22093	75.24	0.000
pH*pH	1	3.53658	3.53658	217.95	0.000
Salt*Salt	1	0.08807	0.08807	5.43	0.034
Temperature*Temperature	1	0.00976	0.00976	0.60	0.450
Time*Time	1	0.00511	0.00511	0.31	0.583
2-Way Interaction	6	0.26962	0.04494	2.77	0.051
pH*Salt	1	0.01448	0.01448	0.89	0.360
pH*Temperature	1	0.11121	0.11121	6.85	0.019*
pH*Time	1	0.13128	0.13128	8.09	0.012*
Salt*Temperature	1	0.00224	0.00224	0.14	0.715
Salt*Time	1	0.00758	.00758	0.47	0.505
Temperature*Time	1	0.00283	0.00283	0.17	0.682
Error	15	0.24340	0 .01623		
Lack-of-Fit	8	0.11164	0 .01396	0.74	0.660
Pure Error	7	0.13176	0.01882		
Total	29	5.68005			

df = *degrees of freedom,* p-*value = probability value, Adj. SS = adjusted sum of squares and Adj. MS = adjusted mean squares. Model Summary: standard deviation (S.D.) = 0.1274, coefficient of determination* (R^2) *= 95.71% and* R^2 *(adj.) = 91.72%.*

Table 2.
Analysis of variance (ANOVA) for CCD based experimental data.

Factors (hyperparameters)	Levels		
	Low: −1	Center: 0	High: +1
Hidden nodes	1	3	5
Batch size	2	6	10
Epochs	5	15	25

Table 3.
Conversion of hyperparameters values permitting to the levels of RSDs.

Weights were initialized by using normal distribution. **Figure 7** demonstrates that there is close agreement between the loss values of training and testing data. From this, it has been generalized that the data is well-trained and there is no over-fitting problem.

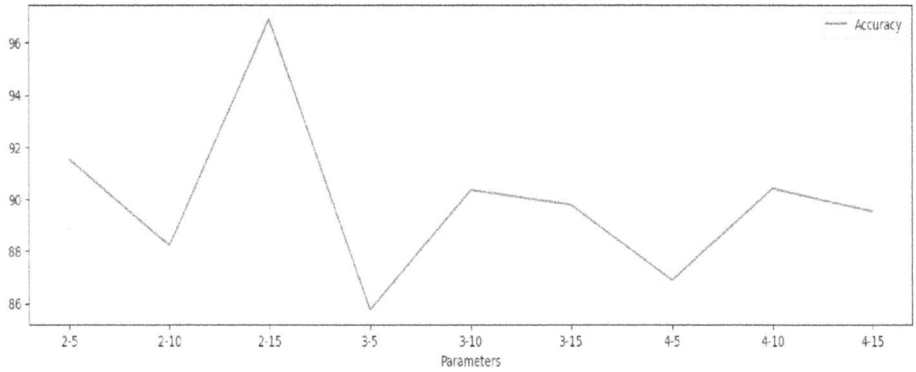

Figure 6.
Optimal value search for batch size and epochs.

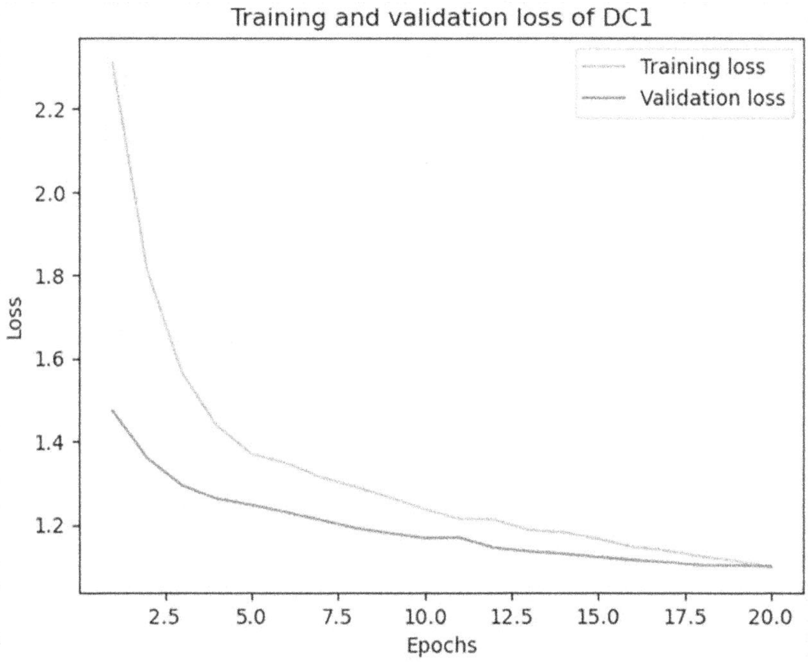

Figure 7.
Model fitting performance for training and testing data.

15. Design structure used to obtain the experimental data

A four-factor CCD was employed to conduct the laboratory experiment. Experimental plan includes 32 trials. The experimental data and fitted data of this subset design are presented in **Table 4**. This design was used to determine the optimized values of the dyeing parameters namely *p*H, salt, temperature and time. The design has provided optimum values by using the given experimental ranges of the color parameters with a less number of experimental trials (**Table 4**).

Design				Response	Estimated response	
*p*H	Salt (g/100 mL)	Temperature (°C)	Time (min)	K/S	RSM-fits	ANN-fits
6	1.5	60	40	2.6501	2.7300	2.6490
8	1.5	60	40	2.5492	2.5043	2.5363
6	2.5	60	40	2.9259	2.8265	2.8993
8	2.5	60	40	2.68	2.7211	2.6777
6	1.5	80	40	2.8491	2.8500	2.8363
8	1.5	80	40	2.3777	2.2908	2.3441
6	2.5	80	40	3.0797	2.9938	3.0775
8	2.5	80	40	2.5688	2.5549	2.5523
6	1.5	60	60	2.4336	2.4505	2.4229
8	1.5	60	60	2.5044	2.5871	2.4960
6	2.5	60	60	2.3762	2.4599	2.3738
8	2.5	60	60	2.7147	2.7168	2.7038
6	1.5	80	60	2.668	2.6237	2.6623
8	1.5	80	60	2.3244	2.4269	2.3320
6	2.5	80	60	2.6325	2.6804	2.6202
8	2.5	80	60	2.687	2.6039	2.6810
7	1	70	50	3.1627	3.1090	3.1606
7	3	70	50	3.3288	3.3824	3.3336
7	2	50	50	3.48	3.3987	3.4848
7	2	90	50	3.3245	3.4057	3.3289
7	2	70	30	3.4348	3.5391	3.4583
7	2	70	70	3.4129	3.3085	3.3981
7	2	70	50	3.3875	3.4803	3.5106
7	2	70	50	3.271	3.4803	3.4421
7	2	70	50	3.4519	3.4803	3.4421
7	2	70	50	3.383	3.4803	3.5106
7	2	70	50	3.5868	3.4803	3.5106
7	2	70	50	3.5625	3.4803	3.4421
7	2	70	50	3.4946	3.4803	3.4421
7	2	70	50	3.7056	3.4803	3.5106

Table 4.
Observed K/S along with fitted values under CCD.

15.1 Prediction comparison of RSM and ANN modeling

Figure 8 shows the performance comparison of RSM and ANN modeling graphically. Observed K/S values were taken on the Y-axis and standard order of

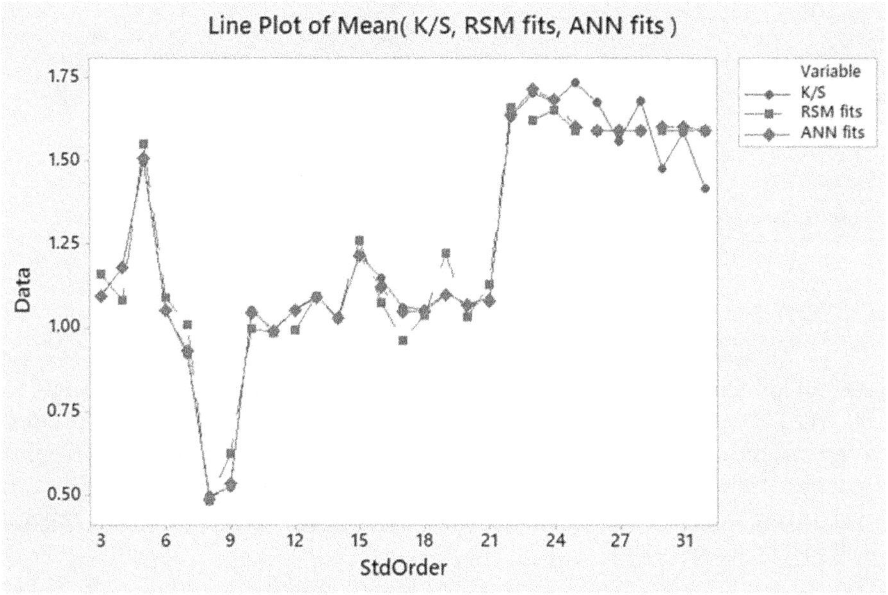

Figure 8.
Observed response K/S and fitted responses for RSM and ANN modeling.

Functions	Inputs
Total number of layers	3
Number of hidden layers	1
Input dyeing parameters	pH, time, temperature and salt
Number of hidden neurons	3
Activation function for hidden layer	ReLU
Activation function for output layer	linear
Optimizer	Stochastic gradient descent
Weight initializer	Normal distribution
Number of iterations	170
Prediction capacity	R^2 and MSE

Table 5.
Architecture of the ANN model.

experimentation on the X-axis. It can be noticed that the estimated data points of ANN are mostly closer to the observed values than RSM. Hence ANN model has shown better prediction capacity than RSM in this particular case (**Table 5**).

By using $\lambda = 0.10$ as a penalty factor that shrinks small weight coefficients to zero for both unconventional and conventional design structures, entries indicated with asterisk became negligible. In **Table 6**, the weight coefficient of the first subscript is for the input and the second subscript is for hidden neuron like: W_{11} is the weight value of the first input and first hidden node; W_{12} is the weight value of the first input

Conventional design structure ($S_4 + S_1 + 8S_0$)		
$W_{11} = 0.1953$	$W_{12} = 0.9894$	$W_{13} = -0.0542^*$
$W_{21} = 0.4668$	$W_{22} = 0.0548^*$	$W_{23} = -0.4242$
$W_{31} = -1.1670$	$W_{32} = 0.1594$	$W_{33} = 0.0815^*$
$W_{41} = 0.0568^*$	$W_{42} = -0.2871$	$W_{43} = 0.3458$

** indicates smaller weight value after imposing penalty factor.*

Table 6.
Weights obtained using LASSO regularization with $\lambda = 0.10$ tuning parameter.

and second hidden node, and same is in rest of the Table. For RSD $S_4 + S_1 + 8S_0$ (CCD); W_{13}, W_{22}, W_{33} and W_{41} coefficients reduce to zero. Therefore, with 10 testing observations, now we have 8 unknown weight coefficients to be estimated from the given design. Hence unknowns can easily be estimated by the given testing data and the over-fitting problem has been resolved. Conclusively the model is now generalizing well for the unseen data.

15.2 Model adequacy using ANN methodology

In the present study, after fine-tuning the over-fitting issue, the ANN model structure with four input factors, three hidden neurons and one response node (4-3-1) has been employed for the given design structures (**Figure 9**). The ReLU was used as a transfer function and stochastic gradient descent (SGD) as an optimizer (**Table 5**). **Figure 8** gives the visualization of the agreement of observed and fitted values for training, testing data.

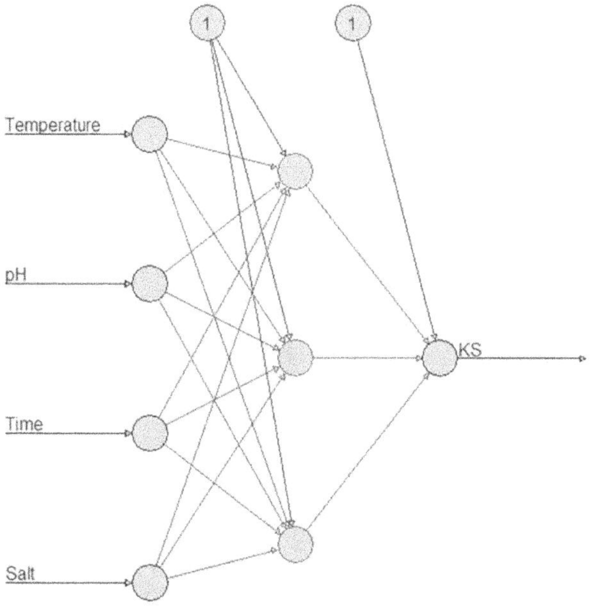

Figure 9.
ANN model structure with four dyeing parameters as input and three hidden neurons.

16. Conclusions

Common scientific experiments include less number of factors. The choice of RSD to fit a particular type of polynomial regression model is a tricky job. The referred design has a flexible experimental region and provides various options to the experimenter to plan his/her experiment. These developed designs for four factors with 32 design points were employed in chemical process i.e. natural dyes. The present study demonstrates the K/S optimization for hulela zard. The important process parameters such as pH, salt, temperature and time of dyeing concentration were used on cotton fabric. $S_4 + S_1 + 8S_0$ (CCD) design has shown excellent performance.

There are also some ML techniques that are recently being implemented for prediction and optimization purposes. For instance, ANN has an amazing ability to derive meaning from complex data and can be used to search patterns and detect trends. Another advantage of using ANN is in the case when the response is not smooth. In this study, ANN-based nonlinear approximation by using an optimal number of hidden neurons is compared with the second order polynomial approximation. The unknown parameters considered were equal in number for the comparison of both RSM and ANN models. Over-fitting was avoided by considering approximation order and unknown parameters.

Traditionally, machine learning hyperparameters were selected through trial and error approaches. In the present work, optimal values of the hyperparameters were determined by using RSDs. Response prediction and optimization is studied by using RSM and ML techniques, especially ANN. The MSE loss function for regression was used to judge the model performance. For the optimization of response, ANN and RSM modeling is generally being used. In literature, most researchers have created a conflict among these models on the basis of end performance as presented in the review section. The researchers have considered very complex models by taking more hidden neurons and also occasionally more hidden layers. Due to such complex models, the predictive performance of the model increases but it causes over-fitting. Due to over-fitting, the model starts to memorize the findings but the objective is a better generalization of the model. RSM is crucial in designing experiments and determining the effects of dyeing parameters, including linear, quadratic, and interaction effects. It is important to note that ANN modeling does not replace RSM, but rather acts as a complementary technique that enhances prediction capabilities.

17. Future directions

After reviewing the whole work presented in this study, the following work in the future can be planned as under:

a. From the view of given material, it is strongly recommended that:

b. Innovative RSDs referred above should also be used for designing purposes in comparison to the conventional design structures.

c. Proper model selection and suitable design plan should be given considerable importance along with ANOVA.

d. Due to the small size study, over-fitting mostly occurs which should also be addressed by using appropriate number of hidden layers and neurons. Sometime we have to use regularization techniques to avoid over-fitting.

e. While comparing prediction performance of RSM and ANN, approximation order should also be considered.

f. For natural dye isolation, this application require such models to make process economic, time and energy saving.

Acknowledgements

The authors acknowledge the use of Copilot for language editing in the chapter.

Author details

Tanvir Ahmad* and Muhammad Aftab
Department of Statistics, Government College University Faisalabad, Pakistan

*Address all correspondence to: dr_tanvir@gcuf.edu.pk

References

[1] Gilmour SG. Response surface designs for experiments in bioprocessing. Biometrics. 2006;**62**(2):323-331

[2] Ahmad T, Gilmour SG. Robustness of subset response surface designs to missing observations. Journal of Statistical Planning and Inference. 2010; **140**(1):92-103

[3] Ahmad T, Gilmour SG, Arshad HM. Comparisons of augmented pairs designs and subset designs. Communications in Statistics-Simulation and Computation. 2018;**49**(7):1898-1921

[4] Box GEP, Wilson KB. On the experimental attainment of optimum conditions. Journal of the Royal Statistical Society. 1951;**13**:1-45

[5] Box GEP, Behnken DW. Some new three level designs for the study of quantitative variables. Technometrics. 1960;**2**:455-475

[6] Mee RW. Optimal three-level designs for response surfaces in spherical experimental regions. Journal of Quality Technology. 2007;**39**:340-354

[7] Victorbabu RR, Ashok D, Narasimham VL. Costruction of second order slope rotatable designs. ProbStat Forum. 2009;**2**:1-7

[8] Montgomery DC, Peck EA, Vining GG. Introduction to Linear Regression Analysis. 3d ed. New York: John Wiley and Sons; 2001

[9] Brownlee J. Statistical Methods for Machine Learning: Discover How to Transform Data into Knowledge with Python. Machine Learning Mastery; 2020. eBOOK

[10] Gonzalez RC, Woods RE. Digital Image Processing, Global Edition. 4th ed. Pearson; 2018. eBOOK

[11] Ertel W. Introduction to Artificial Intelligence. Ravensburg, Germany: Springer; 2018

[12] Perros HG. An Introduction to IoT Analytics. United States (U.S.): CRC Press; 2021

[13] Hecht-Nielsen NR. Kolmogorov's mapping neural network existence theorem. In: IEEE First Annual International Conference on Neural Networks. Vol. 3. New York, NY, USA: IEEE Press; 1987. pp. 11-13

[14] Huang GB. Learning capability and storage capacity of two hidden layer feed-forward networks. IEEE Transactions on Neural Networks. 2003; **14**:274-281

[15] Hornik K, Stinchcombe M, White H. Multilayer feed-forward networks are universal approximators. Neural Networks. 1989;**2**(5):359-366

[16] Berke L, Hajela P. Applications of artificial neural nets in structural mechanics. In: Shape and Layout Optimization of Structural Systems and Optimality Criteria Methods. Vienna: Springer; 1992. pp. 331-348

[17] Carpenter WC, Barthelemy JF. A comparison of polynomial approximations and artificial neural nets as response surfaces. Structural Optimization. 1993;**5**(3):166-174

[18] Sheela KG, Deepa SN. Review on methods to fix number of hidden neurons in neural networks. Mathematical Problems in Engineering. 2013;**2013**(1):425740

[19] Nusrat I, Jang SB. A comparison of regularization techniques in deep neural networks. Symmetry. 2018;**10**(11):648

[20] Baruah J, Chaliha C, Nath BK, Kalita E. Enhancing arsenic sequestration on ameliorated waste molasses nanoadsorbents using response surface methodology and machine-learning frameworks. Environmental Science and Pollution Research. 2021;**28**:11369-11383

[21] Kumar UH, Radhakrishnan P, Shanmugam K, Kushwaha OS. Growth of MWCNTs from azadirachta indica oil for optimization of chromium (VI) removal efficiency using machine learning approach. Environmental Science and Pollution Research. 2022;**29** (23):34841-34860

[22] Singh J, Kumar P, Eid EM, Taher MA, El-Morsy MH, Osman HE, et al. Phytoremediation of nitrogen and phosphorus pollutants from glass industry effluent by using water hyacinth (Eichhornia crassipes (Mart.) Solms): Application of RSM and ANN techniques for experimental optimization. Environmental Science and Pollution Research. 2022;**30**(8): 20590-20600

[23] El-taweel RM, Mohamed N, Alrefaey KA, Husien S, Abdel-Aziz AB, Salim AI, et al. A review of coagulation explaining its definition, mechanism, coagulant types, and optimization models; RSM, and ANN. Current Research in Green and Sustainable Chemistry. 2023;**6**: 100358

[24] Zulfiqar M, Chowdhury S, Omar AA, Siyal AA, Sufian S. Response surface methodology and artificial neural network for remediation of acid orange 7 using TiO_2P25: Optimization and modeling approach. Environmental Science and Pollution Research. 2020;**27**: 34018-34036

[25] Shafer G. Why Should Statisticians be Interested in Artificial Intelligence? 1990

[26] Balkin SD, Lin DK. A neural network approach to response surface methodology. Communications in Statistics Theory and Methods. 2000;**29** (9–10):2215-2227

[27] Johnson RA, Wichern DW. Applied Multivariate Statistical Analysis. Vol. 6. London: UK Pearson; 2014

[28] Myers RH, Montgomery DC, Anderson-Cook CM. Response Surface Methodology: Process and Product Optimization Using Designed Experiments. Hoboken, New Jersey: John Wiley & Sons; 2016

[29] Friedrich S, Antes G, Behr S, Binder H, Brannath W, Dumpert F, et al. Is there a role for statistics in artificial intelligence? Advances in Data Analysis and Classification. 2022;**16**(4):823-846

[30] Agarwal S, Singh AP, Mathur S. Removal of COD and color from textile industrial wastewater using wheat straw activated carbon: An application of response surface and artificial neural network modeling. Environmental Science and Pollution Research. 2023;**30** (14):41073-41094

[31] Cai Y, Xiao L, Ehsan MN, Jiang T, Pervez MN, Lin L, et al. Green penetration dyeing of wool yarn with natural dye mixtures in D5 medium. Journal of Materials Research and Technology. 2023;**25**:6524-6541

[32] Jha AK, Sit N. Comparison of response surface methodology (RSM) and artificial neural network (ANN) modeling for supercritical fluid extraction of phytochemicals from terminalia chebula pulp and optimization using RSM coupled with desirability function (DF) and genetic algorithm (GA) and ANN with GA. Industrial Crops and Products. 2021;**170**: 113769

[33] Kumari S, Verma A, Sharma P, Agarwal S, Rajput VD, Minkina T, et al. Introducing machine learning model to response surface methodology for biosorption of methylene blue dye using *Triticum aestivum* biomass. Scientific Reports. 2023;**13**(1):8574

[34] Rosa JM, Guerhardt F, Júnior RSER, Belan PA, Lima GA, Santana JCC, et al. Modeling and optimization of reactive cotton dyeing using response surface methodology combined with artificial neural network and particle swarm techniques. Clean Technologies and Environmental Policy. 2021;**23**:2357-2367

[35] Salari M, Nikoo MR, Al-Mamun A, Rakhshandehroo GR, Mooselu MG. Optimizing Fenton-like process, homogeneous at neutral pH for ciprofloxacin degradation: Comparing RSM-CCD and ANN-GA. Journal of Environmental Management. 2022;**317**: 115469

[36] Slama HB, Chenari Bouket A, Pourhassan Z, Alenezi FN, Silini A, Cherif-Silini H, et al. Diversity of synthetic dyes from textile industries, discharge impacts and treatment methods. Applied Sciences. 2021;**11**(14): 6255

[37] Vadood M, Haji A. A hybrid artificial intelligence model to predict the color coordinates of polyester fabric dyed with madder natural dye. Expert Systems with Applications. 2022;**193**:116514

[38] Uthayakumar H, Radhakrishnan P, Shanmugam K, Kushwaha OS. Growth of MWCNTs from azadirachta indica oil for optimization of chromium (VI) removal efficiency using machine learning approach. Environmental Science and Pollution Research. 2022;**29** (23):34841-34860

www.ingramcontent.com/pod-product-compliance
Lightning Source LLC
Chambersburg PA
CBHW081334190326
41458CB00018B/5999
9781836346173